Imagining Science

Art, Science, and Social Change

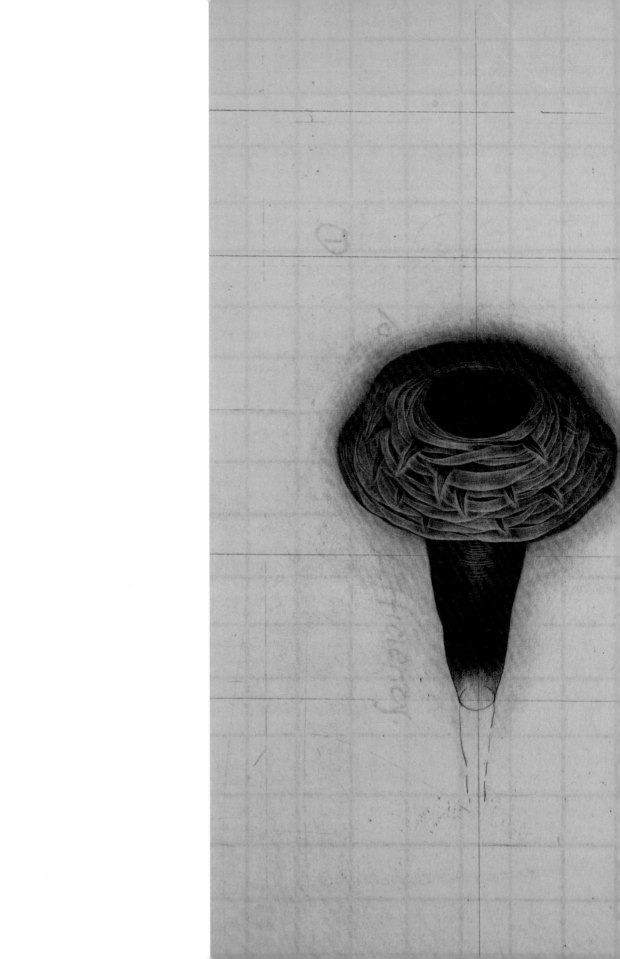

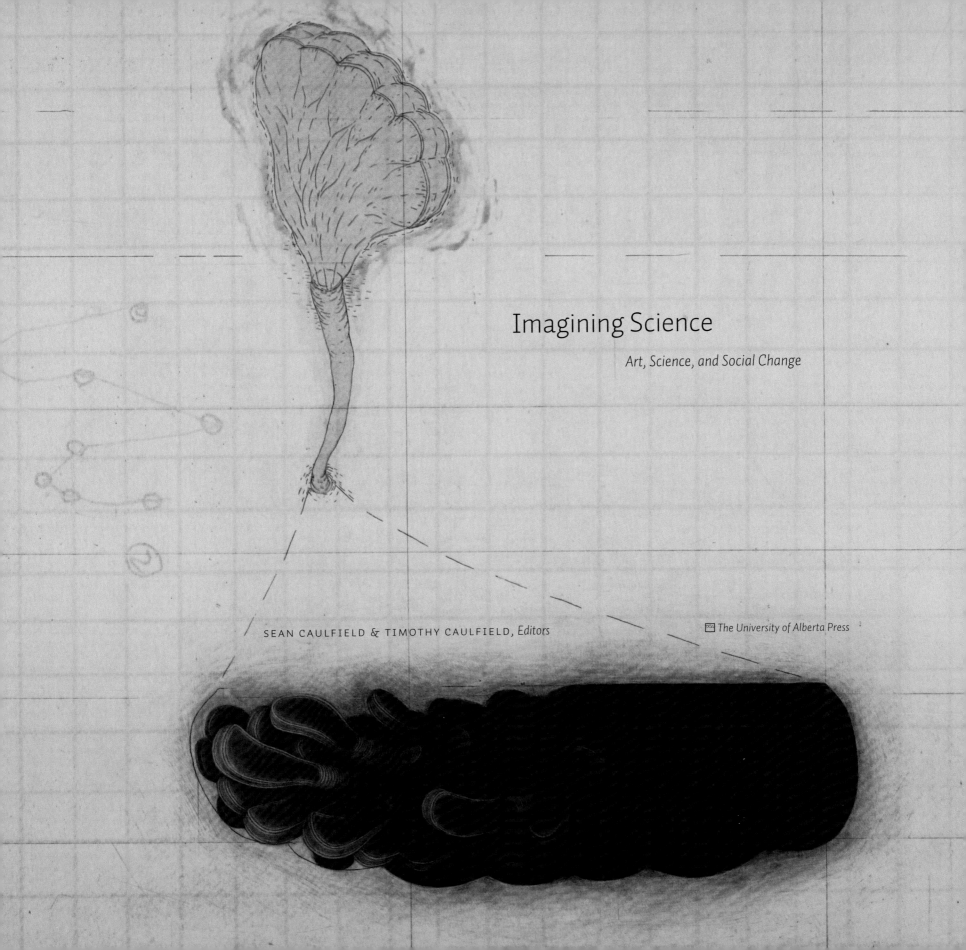

Imagining Science

Art, Science, and Social Change

SEAN CAULFIELD & TIMOTHY CAULFIELD, *Editors*

The University of Alberta Press

Published by

The University of Alberta Press
Ring House 2
Edmonton, Alberta, Canada   T6G 2E1

Copyright © 2008, The University of Alberta Press

LIBRARY AND ARCHIVES CANADA CATALOGUING IN PUBLICATION

　　Imagining science : art, science and social change / Sean Caulfield,
Timothy Caulfield [editors].

ISBN 978-0-88864-508-1

　　1. Biotechnology in art. 2. Art and science. 3. Art—Moral and
ethical aspects. 4. Science—Moral and ethical aspects. I. Caulfield,
Sean, 1967- II. Caulfield, Timothy A., 1963-

N72.B56I43 2008　　701'.05　　C2008-903766-9

The University of Alberta Press gratefully acknowledges the support
received for its publishing program from The Canada Council for the
Arts. The University of Alberta Press also gratefully acknowledges the
financial support of the Government of Canada through the Book
Publishing Industry Development Program (BPIDP) and from the
Alberta Foundation for the Arts for its publishing activities.

Half titlepage image: Lyndal Osborne, ab ovo *(detail), 2008.*
Titlepage image: Sean Caulfield, Body Landscape Diagram *(detail), 2008.*

CONTENTS

FOREWORD

ONE OF THE CHIEF ATTRACTIONS of being a writer and editor is also one of its most humbling realities; namely, that the exhilarating new areas of learning you get to poke around in inevitably break wide open to reveal the vast intricate cavern of your own ignorance. For the last ten years or so, I've been fortunate enough to travel the world and explore the work of scientists, politicians, business leaders, humanitarians, and artists, among many others. It's been both humbling and terrific fun. However, I can assure the readers of this volume that there has never been a project I've been involved in which has opened up such an exciting new way of looking at the world, while simultaneously making me feel almost reptilian in my relative lack of brain power, at least at the start of the project, anyway. After agreeing to edit this book, I wondered what fresh embarrassment I had gotten myself into. But that changed, and here's why.

The essays and works of art in *Imaging Science* are drawn from thinkers and artists of the highest order, as you'll see when you dip into the book. But they are also gifted with the ability to think across borders, borders that inhibit or intimidate most of us (though they shouldn't). Working with the essayists in this book was one of the most enjoyable projects I've been involved in for years, precisely because of their willingness to make it new, to think not just about what they do, but about the meaning of what they do and the ability of an audience to interpret what they do. They did not accept the status quo of writing for a tiny audience of ivory tower residents. They wanted to communicate with a wider audience, a general audience, and it worked for two simple reasons. First, they put their minds to the task because they themselves are so genuinely excited by their work. Second, the work is truly thrilling and they want the world to know about it. The marriage of intellect and enthusiasm in this book is palpable and infectious, and what these thinkers and artists achieved, finally, was to create a cluster of work that is both highly complex while being wholly accessible. This is no small accomplishment.

And that's why you'll enjoy the experience of reading this book as much as I enjoyed working on it. To the contributors, I offer my thanks. To the readers, I offer the promise that in sitting down with this book you have many hours of pleasure and stimulation ahead of you.

PREFACE

CATHERINE CROWSTON, *Deputy Director/Chief Curator, Art Gallery of Alberta*

IN THE FALL OF 2006, Professors Sean and Tim Caulfield, Canada Research Chairs in the fields of Printmaking and Health Law and Policy at the University of Alberta, approached the Art Gallery of Alberta about a unique idea they had for an exhibition and publication to celebrate the University of Alberta's centenary. The project proposed an innovative collaboration between artists, scientists, cultural theorists, bio-ethicists and social commentators as a way to investigate the wide range of issues that have arisen from new developments in the life sciences, as well as the impact they have had on our beliefs, assumptions and values. This type of project, with its cross-disciplinary approach, focus on the creation of new works of art, and its basis in local and international partnerships, resonates with the Art Gallery of Alberta's core mandate and our belief that the inspiration for art comes from all areas of contemporary life.

The artists that we invited to participate in this project represent a diversity of aesthetic practices. While some have been critically engaged with an examination of science and bio-technology for many years, others came to the project with a sense of wonder and excitement, responding to our challenge to take their work in new directions. What has resulted is a multi-layered exhibition and publication with a diversity of approaches to the subject at hand, with works of art that visually reference the objects of science: its tools, devices and hardware;

works that reference the human body and the mysteries of its form and interiority; works that critique specific bio-technical developments, such as genetic modification and gene alteration, and works that directly engage with biotechnology's accepted methodologies and protocols.

While the final works vary in the level of their engagement with the specificities of science and its trajectories, inevitably none of them are representations of science itself. They do not provide explanations or justifications for science and its infinite areas of research and investigation. Instead they are reflections, critiques, musings and semblances. They are not about fact per se, but about experience, interpretation and contemplation. They engage us with the possibility of their ideas and images, and leave us to decide what we will learn from them.

ACKNOWLEDGEMENTS

WE WOULD LIKE TO THANK the University of Alberta's Centenary Project Fund, Killam Research Fund, President's Grant for the Creative and Performing Arts, and Faculty of Law, in addition to the Art Gallery of Alberta, Genome Alberta, the Social Sciences and Humanities Research Council, and the Alberta Heritage Foundation for Medical Research for their generous support of the *Imagining Science: Art, Science, and Social Change* workshop, exhibition, and publication. Individually, we would like to thank C.J. Murdoch and Nina Hawkins for their help with the organization of the workshop, Curtis Gillespie for his superb editorial work, Catherine Crowston for her organization of the exhibition, and Miki Andrejevic for his help with the co-ordination and funding of the public speakers' panel. We are also enormously grateful to the artists and scholars who agreed to participate in this exciting and unique project and to The University of Alberta Press for agreeing to publish this book under such strict timelines. Lastly, we save a special thanks to Robyn Hyde-Lay for her assistance with all aspects of the project. Without her energy, enthusiasm, and skills, this project would not have been a success.

Beauvais Lyons

"L'Anatomie" Plamche 144, 2004

Lithograph

Courtesy of the Hokes Archives

Photo: Gary Heatherly

Image from "Hokes Medical Arts,"

an exhibition of prints and drawings

from the Hokes Archives that is a work

of academic parody.

TIMOTHY CAULFIELD

WE LIVE IN A TIME that has been characterized by many as the start of the biotechnology century—an era during which biotechnology will emerge as the dominant scientific field and influence a broad array of human endeavours. Even if one does not agree with such a grand prediction, it is hard to deny that biotechnology has become, rightly or not, a focal point of research investment, scientific inquiry, economic activity and public debate. Indeed, emerging areas of biotechnology, including genetics and stem cell research, have the potential to drastically alter the way we treat disease, view ourselves and our communities, and conceive children. Biotechnology has permeated the landscape of popular culture. It can be found on TV, in Hollywood movies and throughout the mainstream press. Rarely a day goes by without at least one headline about biotechnology—be it a new breakthrough or a new controversy.

In many ways, it is the profoundly controversial nature of the field that makes it so unique, at least from the perspective of social policy. The great potential for benefit is accompanied by possible risk. Public excitement about promising medical breakthroughs is often accompanied by uncertainty about the ethical appropriateness of the research. So, as the research moves forward, we continue to struggle with a surplus of seemingly unresolvable ethical dilemmas. Does the human embryo have sufficient moral status to preclude stem cell research? Should we use genetic testing to select the traits of our offspring? Should we be able to patent genes? Will knowledge of our genetic makeup lead to an overly deterministic view of humanity? Such questions, and many more, have been debated for decades and remain a focal point of academic inquiry for a wide variety of scholars, including artists.

Art has always been a powerful tool of social commentary. In the West, Pre-Renaissance art was often used as a way to legitimize and glorify the status quo and the ruling icons, be they secular or spiritual. Subversive elements snuck in, but art's most visible social utility was not as an agent of change. During the Renaissance, the palette expanded greatly. The economic, philosophical and scientific changes that were occurring throughout Europe were reflected in (and, no doubt, partly caused by) new perspectives in art. In the 1800s, art challenged social norms and the conventions of artistic expression.

Today, art continues to play the role of provocateur. Many present-day artistic responses to the fundamental ethical questions associated with biotechnology build on past creative works, such as Leonardo da Vinci's portrayals of the human form and contemplations on the status of the human embryo. However, the pace and scope of change that society is facing today as a result of scientific advance is so pervasive that the specific, targeted and truly unique field of "bio art" has arisen. Bio art is a field of artistic inquiry that both utilizes the techniques of biotechnology and serves as a medium of reflection on the societal implications of the research.

As a mode of science commentary, art can be both commanding and subtle. Unlike an academic article, a quick glance at a piece of artwork can produce an immediate response in its audience—be it wonder or revulsion. It can draw us into novel situations or invite us to see things from a fresh perspective. But artistic images may also affect the viewer on a more understated level by simply asking us to see beauty in new places and in new ways.

In the context of biotechnology, art is used to put forward a gamut of perspectives. Bio-conservatives voice concern about the impact of biotechnology on our environment, health and personhood. Some of this work has the force of a broad intuitive reaction to the entire area of research. Here, the theme is often one of caution. Other work is more specific in its intent, tackling discrete topics, such as the role of market forces in shaping a research agenda. But other work suggests a less pessimistic outlook. This art embraces the potentially transforming power of biotechnology, asking the viewer to consider "what if?" and, perhaps more provocatively, "why not?"

This book includes a unique collection of artwork, all of which has been inspired and informed by emerging areas of biotechnology. The works echo the continuum of viewpoints referred to above. But this book is much more than a collection of art with a similar theme. We sought to bring together artists, scientists and social commentators from a wide variety of disciplines to share perspectives on biotechnology and the interplay between art and science. Our goal was to provide a wide range of commentary—scientific, philosophical and legal—and place art in the mix. To this end, we also collected short essays from renowned scientists, journalists, philosophers, legal scholars and social science experts.

The work represented in this text also reflects a collaborative, creative and truly interdisciplinary approach. All of the book's artists and many of the essayists participated in an exceptional workshop at the Banff Centre, in Banff, Alberta, in the summer of 2007. Here, we shared ideas, discussed emerging social concerns and engaged in a truly open exchange of perspectives. The dialogue informed both the development of this book and the plans for the associated show at the Art Galley of Alberta (November 14, 2008-February 1, 2009).

In editing this book, we put few constraints on either the artists or the essayists, as long as they touched on the broad themes described above. We wanted pieces that were provocative, engaging and that would encourage readers to consider the social issues associated with biotechnology in a new light. Our contributing colleagues did not disappoint. The essays cover everything from the nature of science and art to a description of the human body as an ecosystem. Some essays explore the parallels between the work of artists and scientists, and others grapple with a specific social issue, such as the creation of human chimeras.

The artist contributions are equally broad in scope, both in theme and approach. Like their colleagues in other disciplines, these contemporary artists are exploring complex questions by working in an increasingly interdisciplinary manner. Their work comes from a wide range of approaches, including utilizing traditional materials and methodologies, as well as alternative creative practices, such as installation and performance.

We hope this eclectic compilation of prints, paintings, sculptures, installation pieces, essays, poems and even song lyrics provides the reader with a unique and thought-provoking taste of the varied perspectives on the social implications of biotechnology.

As co-editors of this book, my brother, Sean, and I were humbled by and grateful for the efforts of everyone involved with this project. Whether scientist, poet, artist, philosopher or lawyer, all participants brought inspiring enthusiasm and creativity to the task at hand. There seemed to be recognition that this was a unique opportunity to explore the relevant issues from a new angle and with fresh input from outside our usual professional circles. We believe that this wonderful collaborative spirit is reflected in the creative works that fill the pages of this book.

LORI ANDREWS & JOAN ABRAHAMSON

WHEN THE U.S. President's Council of Advisors on Science and Technology decided to address the issue of nanotechnology, the council members began by reading Michael Crichton's *Prey*. When the U.S. Department of Defense's Defense Advanced Research Projects Agency (DARPA) sought fresh ideas about the biotechnological future, they empanelled a group of novelists, including science fiction author Nancy Kress. When the U.S. Congress addressed reproductive technologies, members referred to Margaret Atwood's *A Handmaid's Tale*. When President George W. Bush addressed the nation about embryonic stem cells, he referred to Aldous Huxley's *Brave New World*.

Art and literature have an effect on the formation of legal policy. After Dorothea Lange's photo, *Migrant Mother*, appeared in the *San Francisco News* in 1936, the U.S. government allocated $200,000 to establish a migrant camp for homeless workers. In a project funded by the Greenwall Foundation, we systematically reviewed hundreds of novels, short stories, representational artworks inspired by genetics and "wet works"—artwork employing techniques (such as tissue culture and gene cloning) borrowed from life sciences. We learned that artists, even more than scientists, can make a contribution to the policy surrounding the life sciences.

By and large, artists (including representational artists, life science artists, novelists and poets) are knowledgeable about the scientific underpinnings of their work. Novelists wrote about in vitro fertilization, the Internet, the atomic bomb and many other technologies before they were developed. In fact, when Cleve Cartmill published the short story "Deadline" in the March 1944 edition of *Astounding*, Allied counterintelligence operatives immediately jumped to the conclusion that someone had leaked information to him about the secret project to develop the bomb. He had merely extrapolated from the existing scientific articles of the times.

In the fine arts, artists are learning the techniques of genetic testing, recombinant DNA, and tissue culture, and incorporating them into their work. Art students at the University of Western Australia use life science techniques in their work to gain masters degrees in Biological Arts granted by the university as a science degree. At MIT, Joe Davis has been an artist-in-residence for over a decade and has created ways to encode messages in DNA.

Scientists and artists can learn from each other. There is a commonality of creativity, even if it is expressed in different ways. When the National Science Foundation prepared a report on the converging technologies of genetics, computers and nanotechnologies, it included many speculative technologies that did not yet exist, based on the inspiration of science fiction writers. Artist Adam Zaretsky showed that tissue cultures grow more rapidly when exposed to music (Englebert Humperdinck, of all things; classical music had no effect). Artists Oron Catts and Ionat Zurr created tissue culture techniques that are now used by biotech companies.

Artists (from representational artists to fiction writers) can provide a map to the personal and societal issues a new scientific development will raise. Artists who are inspired by genetics are able to focus attention on issues about the social and personal impacts of biotechnologies. Artists highlight political issues regarding the unpredictability of outcomes, the incentives within science and society for pursuing particular technologies, and the power struggles within society that will influence how the technologies are implemented. In the personal realm, artists focus attention on the risks and benefits of the technologies, the distributive justice aspects of technology transfer, and the impact of the technologies on self-concept, stigma, rewards, culture and aesthetics.

New ways should be developed to incorporate the thinking of artists in the policy sphere. When the Human Genome Project was initiated, proponents described it as "biology's moon shot," while detractors thought of it as "the Manhattan Project of Science." The development of genetic technologies raises important social issues. Should an insurer or employer be able to require genetic tests and reject applicants who have a predisposition to a genetic disease that is costly to treat? Should parents be able to (or forbidden to or required to) use genetic technologies to "enhance" their embryos? What incentives are necessary to encourage appropriate

genetic research? Should the sequence of a gene—the basic scientific alphabet of A, C, G, T—be owned? These questions are thought to be so important that the publicly-funded genome initiatives around the world grant funding to social scientists and philosophers for research into the ethical, legal and social implications of genetics.

But perhaps an alternative way to explore the ethical, legal and social impact of technologies like genetics is to analyze the issues artists predict will arise with the technologies. After all, as the French poet, artist and director Jean Cocteau noted, "Art is science made clear."

THE EVOCATIVE IMAGE

Art and the Perception of Risk

CONRAD BRUNK

ART CAN REVEAL aspects of a technology that remain hidden to the uncritical eye. Technology has ways of hiding its own implications and hidden meanings. As the philosopher Martin Heidegger has famously pointed out, technology has a way of "framing" the world for us in ways that reveal certain aspects of itself and the world, while hiding others. Good art can "reframe" that technologically framed world and sometimes even "unframe" it.

One of the ways art can do this is through the creation of images that evoke passionately emotional, even visceral, responses to new technologies. It might be argued that the artist is manipulating our acceptance or rejection of a technology by shaping it in ways that bring extraneous negative or positive associations into play. But, it might also be argued that the artist is drawing out potential, or even intrinsic, aspects of the technology that challenge deep-seated and profoundly authentic moral or spiritual values. Surely good art can do both these things, but when it aims at the "truth," it can effectively do the latter.

Art rooted in exploring biotechnology can perform this truth-revealing role. It can do so by evoking strong "gut" reactions, though this necessarily prompts the question of what is behind the reaction itself. Or, more precisely, it poses a question long debated by moral theorists: Are the moral judgements we make at the level of rational discourse independent of (or even in opposition to) our visceral, emotional reactions, or are those moral judgements integrally bound up with our emotions? Immanuel Kant epitomizes the first view, David Hume the second.

Artists working with biotechnological themes persuade me to adopt the second view. They do so in the way they call attention to possibilities within the new genomic technologies that give us pause as well as enthusiasm. Their immediate appeal is to our emotions, perhaps, but the reflection on *why* we have the emotions we do when confronted with the images is just as important. My view is that deep visceral emotional responses often are the expressions of basic underlying moral sentiments that can be made explicit and defensible within a framework of moral rationality.

The science of risk analysis has become the dominant decision-making tool of technological societies when dealing with threats to health and environment. Within this science "risk" is understood as the *probability* of a *harm* occurring as a result of a human activity, a technology or a natural process (with the attendant assumption we now carry that technology helps us control consequences). Protecting health and safety while developing technologies is primarily a problem of finding the technologies that reduce risks to a minimum while also realizing the maximum benefit from the risk taking. Technologies are science based, and science requires the reduction of every problem to a quantifiable algorithm. So, the benefits and harms, as well as their probabilities, that "really count" are only those easily definable in quantifiable terms. Ordinary people, however, usually don't think this way about safety issues. They find many risks to be acceptable when the risk analysts consider them highly unsafe—activities such as sky-diving, driving automobiles and unprotected sex. Conversely, many people often find many risks unacceptable that the risk experts consider acceptably safe—like pesticides on their foods and lawns, radioactive waste buried under their backyards, and genetically modified foods. The experts say that ordinary people simply fail to understand or appreciate probabilities, overestimating the risks of extremely improbable adverse events and underestimating the risks of highly probable ones. They focus too much attention on the potential *consequences* of a risk and too little attention on its *probability*.

Studies of risk perception identify multiple factors that significantly influence ordinary people's assessment of risks and their level of acceptance of these risks. It turns out that most people pay far closer attention to the qualitative aspects of risks that make them *acceptable* or *unacceptable* than they do to the simple issues of the magnitude of the risk and magnitude of the potential harm. The question of the acceptability of risk (the definition of "safe" in risk analysis) is complex since acceptability involves deep-seated, often strongly contested, emotional and moral responses to the risk. For example, the risk of a harm you freely take upon yourself is, by definition, more acceptable to you than one imposed upon you by another. Similarly, the risk of a harm

to you or to others that results from what you hold as the unethical activity of another is not likely be acceptable (therefore not "safe") to you. Risks of harms that people find "dreadful" in themselves will be less acceptable, even though less probable, than those that do not carry such dread. Death by cancer, Creutzfeld-Jacob disease, and by conflagration are examples of such harms.

What the artist can do so effectively, as illustrated by those artists inspired by biotechnology, is to call our attention to these underlying complex moral sympathies that are often the most important factors in our willingness to accept a technology as "safe" (ethically acceptable). They do so by highlighting those aspects of the technology that evoke these sympathetic responses.

Jason Phillip Knight &
Fiona Annis
"for anna (death in labour)"
1926–28 and 2007–08
Digital print (original destroyed)
Courtesy of the artists

for anna knight

there is no shame

linda (unspoken)
there were two families

never fell – but he

broke his leg – gave his name to another sister – clippings
they later built a tower

an unknown date – high on hay bales
lost there too – 16 & 13

it happens – orchidea acer erables

anna – marked she disappeared
bore child

she died unknown

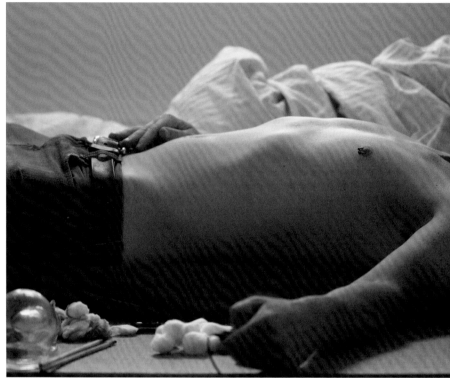

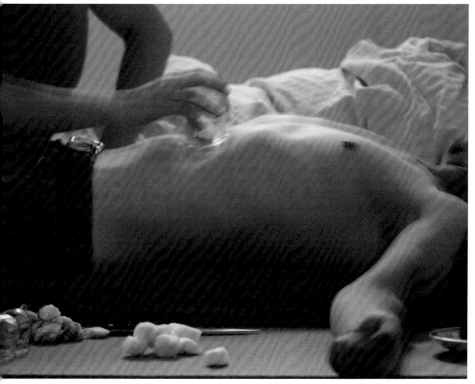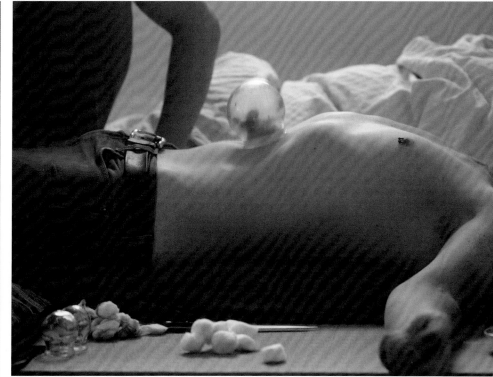

Much of the dialogue related to bio-ethics, reproduction, and medicine can occur at a very broad and highly theoretical level. As important as this discourse is, it is also crucial to examine issues raised by these fields from a more human dimension, exploring the psychological and emotional dimensions that arise when technology, ethics, public policy and social/cultural norms intersect with individual needs and desires. *Corpscellule*, a collaboration between Shawn Bailey and Fiona Annis, uses digital images and text to create an evocative, poetic and highly personal vision in which the audience is invited to examine the interplay between social and political forces with relationships, genealogies of illegitimacies and personal tragedy. The piece is part of an ongoing cycle of work that explores the intersections of the sublime and the grotesque, unravelling the peripheries of revelation and concealment, employing the body variously as material, site and agent.

HALF HUMAN, HALF BEAST?

Creating Chimeras in Stem Cell Research

CYNTHIA B. COHEN

WE ARE ALL FAMILIAR with the Chimera, which has emerged from ancient Greek mythology and been so often represented in works of art—a creature with the front parts of a lion, a goatlike middle, and the hindquarters of a snake. Stem cell scientists now tell us that it is essential to develop human-nonhuman chimeras—living beings composed of both human and nonhuman cells—in order to pursue stem cell research. They point out that stem cells, particularly embryonic stem cells, have great medical potential, since they can be transformed into almost any cell of the body. One day differentiated human stem cells might be transplanted into patients with cardiac disease and into children suffering from diabetes to treat and even to cure them. At this time, however, we don't know how these flexible cells might behave once inserted into the bodies of humans; it could be extremely unsafe. Consequently, stem cell scientists have found a need to study the ways in which they spread and differentiate in animals first. Doing so, however, inevitably results in the creation of human-nonhuman chimeras.

A basic difficulty with the development of such chimeras is that we don't know how much the transplanted human stem cells will contribute to the brains of the nonhumans receiving them. If such cells were injected into prenatal animals, especially at a very early stage, would this result in animals with human brains? If so, might these animals think and act like humans?

Such concerns were raised when a team at Stanford University proposed suffusing the brains of newborn mice with human neural stem cells in hopes of learning how to reconstruct damaged brain tissue. The group came under fire from some other scientists who saw their proposed experiment as a risky endeavour that would raise the ethical stakes. Media reports envisioned the birth of curious creatures that looked like mice but behaved like humans. They again raised the spectre of the fire-breathing mythological chimera, this creature that left death and destruction in its wake. Contemporary human-nonhuman chimeras struck them as equally sinister.

Can we put our collective finger on why we tend to draw back from the idea of creating human-nonhuman chimeras?

Essentially we are concerned that certain distinct and valuable human capacities would be transferred to animals in a diminished form were we to create nonhumans—particularly primates—with human brains. Doing so could violate human dignity, which is the catch-all concept referring to a family of capacities distinctive to humans that include the ability to reason, make choices, carry out moral evaluations, speak in a complex language, participate in interweaving social relations, sympathize with others, and so on. The human brain is closely associated with the family of capacities especially relevant to human dignity. If chimeras with human brains were created in the course of stem cell research, the physical limitations posed by their nonhuman bodies could diminish the way in which those human brains functioned, and, as a result, also diminish those human-dignity-related capacities. It would therefore be wrong to develop such human-nonhuman chimeras, many argue.

This does not mean, however, that all research involving the insertion of human embryonic stem cells into prenatal nonhumans should cease on ethical grounds. In 2005, a team headed by Fred Gage at the Salk Institute in La Jolla, California, reported that they had injected 100,000 human embryonic stem cells into the brain ventricles of embryonic mice. (Mouse brains usually contain 75 million to 90 million mouse cells, so this was a relatively small number.) They did so in an attempt to make a realistic model of neurological disorders, such as Parkinson's disease, and eventually to develop treatments for such conditions.

The human embryonic stem cells differentiated into human neurons and integrated with the mouse neural cells in the adult mouse forebrain. These human neurons were the same size as the mouse brain cells and responded to mouse brain signals. They composed only about 0.1 per cent of the cells in the mouse brains. The researchers concluded that human embryonic stem cells inserted into the brains of prenatal mice in relatively small amounts had not changed the way in which the mice functioned. The human cells matured into the kind of cells that surrounded them and were controlled by the brains of the mice. This study did not result in humanized mice that could drive cars or read poetry.

However, less is known about what would result if human cells were inserted into nonhuman embryos at their very earliest stages. Here the potential for creating human-nonhuman chimeras with human brains seems strongest, for the brain is just beginning to develop.

To address this issue, the guidelines of the Canadian Stem Cell Oversight Committee prohibit funding research involving the insertion of human or nonhuman embryonic stem cells into human or nonhuman embryos or fetuses. This prohibition seems overly broad, however, in light of the results of the Gage study, in which human embryonic stem cells were inserted into late-stage nonhuman embryos without ill effects. For this reason, the National Academy of Sciences in the United States maintains in its voluntary guidelines for the pursuit of human embryonic stem cell research that researchers should not insert human embryonic stem cells into nonhuman embryonic primates at the blastocyst stage, which is very early. Because primates and humans are remarkably close in their developmental potential, there is particular concern about creating a half-human half-primate being. In contrast, the Science and Technology Select Committee of the Parliament of the United Kingdom advised in 2007 that the regulatory authority should develop rules for the transfer of human stem cells into animal blastocysts as a condition of licensing stem cell researchers seeking to pursue such research. This body apparently believed that it could decrease the possibility of creating nonhumans with human brains through careful regulation.

The mingling of bodily materials of humans and nonhumans has become more acceptable since the ancients first developed the notion of the monstrous chimera. What is at issue today, however, is whether the transfer of human stem cells to nonhuman embryos would result in the creation of human-nonhuman chimeras that violate human dignity. Several countries have drawn the basic conclusion that such studies need not threaten the belief at the core of our social ethic that human beings have a certain distinctive dignity, but the studies could, instead, if conducted according to carefully developed ethical and scientific guidelines, uphold that central conviction about human dignity and at the same time promote human well-being.

FRANCIS S. COLLINS

I GREW UP IN A HOME where music played a central role. My father was classically trained on the violin, but he had been a folksong collector in North Carolina in the 1930s, so he knew that a violin could also be a fiddle. My mother sang haunting Elizabethan songs as he accompanied her on the guitar. Later my father became an international authority on medieval music dramas, transcribing and bringing to life numerous religious plays that had existed for hundreds of years only as dusty manuscripts in European abbeys. So it was natural, as I was growing up in the Shenandoah Valley, to be surrounded by music and musicians. The sounds that filled our farmhouse on any given evening could include everything from Archangelo Corelli to Bill Munroe.

I thought about music as a career, but science grabbed me in the tenth grade and wouldn't let me go—first chemistry and physics, later biology, genetics and medicine. But the connection between science and music was immediately apparent. After all, the reason that certain musical chords sound pleasing to our ears, and others sound dissonant, relates to the physics of sound waves—if the frequencies have integer relationships to each other, the overlapping synchronized peaks of sound hitting your eardrums creates a pleasant experience.

Ultimately my scientific interests coalesced around the study of DNA—this marvelously simple double helix that serves as the information molecule of all living things. As a digital method of storing information, protecting it from damage, and providing a means of accurate copying, DNA is incredibly elegant. After working with this molecule every day for the last three decades, I am still in awe. And that awe is not just mathematical or coldly scientific, it is numinous. It enters the same mental space that is filled by an exquisite sunset or the playing of Mozart's Piano Concerto No. 23.

So can we make some sort of direct connection of DNA to music? Well, DNA has only four letters in its alphabet (A, C, T, G)—four tones in its scale, one might say. That's a little short of the eight- or twelve-tone scale (but at least it's part of the same arithmetic series). Some have tried to turn the DNA sequence of human genes or chromosomes into music by some sort of rule about how to convert A, C, T, and G to Do, Re, Mi, Fa, So, La, Ti, Do, but it always comes out a little forced. I think I should like it, but I can't quite make the leap. Part of the problem, of course, is that DNA musical transcribers don't know the rhythm—just reading the letters of DNA at a standard pace leads to an intrinsically boring melodic line, because there's no variation in the intervals between notes. That's probably not the way our own cells read DNA—no doubt the RNA polymerase that reads the genes and starts the process of turning that information into protein has slow movements and fast movements. DNA's got rhythm, we just haven't been able to learn its time signatures yet.

Come to think of it, this was just the problem my father faced when he was trying to bring those medieval manuscripts to life.

You see, musical notation in the thirteenth century only included pitch, it didn't tell you the rhythm. Was it 4/4 or 3/4? Was that a quarter note or an eighth note? There was no way to tell. Because of this ambiguity, stuffy musicologists relegated these exquisite manuscripts to the academic dustbin—but my dad learned everything he could from the oral tradition of the troubadours about the musical idioms of the time and then gave it his best shot. Listening to the haunting strains of his production of the *Visitatio Sepulchri*, echoing through the Abbey St. Benoit de Fleury, I had to conclude that he did the right thing.

Are we genome scientists so different? We have the one-dimensional script of the human genome (the notes), and we're trying to give our understanding of DNA's function (the rhythm) our best shot right now. What we produce is not a completely accurate rendition, but it guides and inspires us just the same.

By now you're wondering where this is all going. Well, maybe so am I. So to leave you with something a little less mythical, I have one other desirable connection between DNA and music to propose. Down through history, we have celebrated or mourned significant moments in song. Music lifts us to a higher plane and joins us together in ways that words alone cannot do. And so, when scientists, ethicists, families affected by genetic illnesses,

and just plain folks gathered a few years ago at the Smithsonian Institution to note the arrival of the sequence of the human genome, it was music that called us all together. Modifying the words of a familiar folk song called "All the Good People" (originally written by Ken Hicks), I had the great privilege of leading these good people in a song of gratitude and hope:

THIS COMMON THREAD

Chorus: *This is a song for all the good people*
All the good people whose genome we celebrate
This is a song for all the good people
We're joined together by this common thread.

This is a song for all of those dreamers
Who're looking for answers to come our way
Families, doctors, researchers, all seekers
We share in the hope for a brighter day.

This is a song for those leaders of science
Who worked in six countries, including the USA,
Germany, China, Japan, France, and England
They worked without resting and gave it away.

This is a song for the ethical experts
Who see the bright promise but also the fears
They press for awareness, for justice, for fairness,
They're part of our conscience, our eyes, and our ears.

This is a song for those who have suffered
Your strength and your spirit have touched one and all
It's your dedication that's our inspiration
Because of your courage, you help us stand tall.

It's a book of instructions, a record of history
A medical textbook, it's all these entwined
It's of the people, it's by the people,
It's for the people, it's ours for all time.

In music, in art, or in science, my father, myself, all us restless humans, we are all driven by curiosity, by a hope to make the world a better place, by a search for God, and yes, by a longing for beauty. One might even say those characteristics are what make us human. May we continue to explore them—all of them.

Adam Zaretsky

SubHuman

Transgenic Pheasant Embryological
Arts Lab, Leiden University Honours
Program, 2007.
Courtesy of VivoArts
Photo: Jennifer Willet

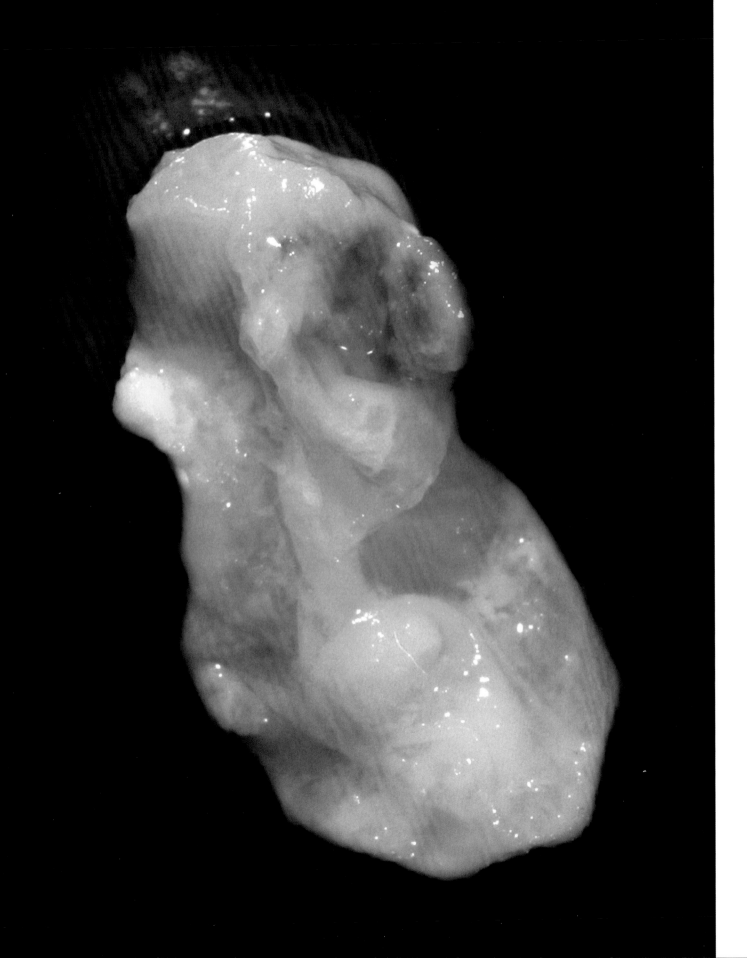

Adam Zaretsky

Ancestor Rape

Transgenic Pheasant Embryological
Arts Lab, Leiden University Honours
Program, 2007.
Courtesy of VivoArts
Photo: Jennifer Willet

Adam Zaretsky

Heritable Palette

Transgenic Pheasant Embryological
Arts Lab, Leiden University Honours
Program, 2007.
Courtesy of VivoArts
Photo: Jennifer Willet

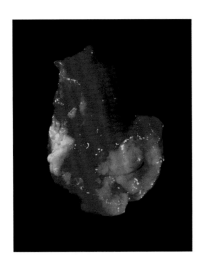

Birdland

Avian Developmental Embryology Arts Project

These are documents from the Transgenic Pheasant Embryology Art and Science Laboratory taught by Adam Zaretsky at the University of Leiden as a part of an honours course called Vivoarts Art and Biology Studio. Held at the University of Leiden, this lab cleared the Animal Experimentation Committee and Recombinant Safety, since it appears that by definition in the EU, avian embryos are not considered animals as they are not "free living beings." They are also not considered a recombinant safety hazard because they are incapable of reproducing. Held by the Arts and Genomics Centre, this hands-on perfomance art wet-lab was documented in order to stimulate debate about the use of new biological methods for permanent alteration of genetic inheritance.

Transgenic Sculpture Laboratories give humanities students the tools and skills they need to implicate them- selves in a real biotechnical relationship with their four-day incubated and windowed eggs. Non-specialist students are offered microsurgical, teratological and naked plasmid injection as developmental embryology tinkering tools, and given a chance to make their first transgenic vertebrate, an embryonic pheasant.

Without expensive equipment for this lab, the students and I tested whether manual, intramuscular plasmid injec- tion with homemade microinjectors could be a technique of pheasant genetic modification. A plasmid is a circular ring of DNA that can unfurl and insert its gene load into a living organism's cells. Secure in the knowledge that the Hepatitis B vaccine is working example of raw plasmid injection, the success of our research echoed the fact that plasmids need not be forced to invade their nuclear targets. Plasmids can find their own way through pores in cellular membranes and incorporate into genomes without the more brutish and costly transgenesis skill sets (i.e., intranuclear microinjec- tion, biolistics, viral vectors, heat shock, electroporation, etc.). When it comes to infectious genes, proximity is some- times all that is needed to alter heredity.

This type of lab furthers qualitative knowledge through hands-on artistic avian embryology and mutagen- esis research. The documents you see are the embodied protocol and the results of imaging somatic difference. The habit of inserting an engineered plasmid into the genome of a cell line or organism is a physical artifact that stems from the mortal desire for lasting signature. These stillborn sculptures, in accord with the libidinal economy of multi- generational directionality, have been impressed upon for the record alone. Consider their mutations to be a sort of genetic graffiti.

In the name of transgenic art, fledgling artists are using lab technique as a new medium to produce living and often mutant living art forms. As their "sculptures" live and die, often at the whims of the artistic investigator, the personal, non-repeatable moments take on a ritual air. What kinds of rituals do interdisciplinary art and biology practices entail? How do they reveal the implicit rituals of science? What new performative rites come out of mixing ethics and aesthetics in the laboratory? Scientific methodology has its own aesthetic which mixes creative flourish with humane sacrifice. But readings of the difference between scientific and artistic play is often based on paradigmatic alterity concerning what the act of experimentation is? As artists learn laboratory technique, the rituals of science and new rituals of sci-art unfold, decouple and reconfirm magical thinking in both arenas.

How does animal research relate to the history of animal sacrifice? What is the role of subjectivity in developmental embryology? Is transgenic protocol also a ritual for the cultural production of liminal monsters? And how does mutagenesis impede or coerce the imaginary in the life- world of heredity? Through an analysis of artists confronted with the responsibility of culling a transgenic pheasant embryo—which they had imprinted with plasmids in the name of art—I hope to show living rituals for new biotech- nological processes as they are invented. Along with my pedagogical role, I took the time to apply these experiments in new media to my personal art practice, developing an embryo named Fratricide to her legal limit.

OF LADDERS, TREES AND WEBS

Genomic Tales Through Metaphor

EDNA EINSIEDEL

METAPHORS provide important connections between language, creativity, science, art and publics. Metaphors are common in everyday discourse, are the lifeblood of artistic and literary forms and are key to communicating scientific ideas, as well as formulating scientific problems. Such metaphors as "chaos," the "big bang," the "greenhouse effect," or "genetic information" illustrate links between popular discourse and scientific thinking. The use of metaphoric images has also played an important role in helping society think about and visualize its confrontation with new ideas and to consider ways of acting.

As society begins to grapple with the work of evolutionary genomics, linguistic tools and artists may help us imagine our place in the evolutionary picture and our complex relationships with other organisms in our world. The shifts in metaphorical thinking around genes, organisms and environments illustrate the beauty and power—as well as the limits—of metaphors.

One of the early metaphors that influenced both scientific and public imaginations is that of the "evolutionary ladder." This picture of a ladder has a common incarnation of the evolution of humans from ape-like ancestors. This ladder of life also has microorganisms typically at the bottom while humans are "highest" in the evolutionary chain. Underlying this evolutionary picture is an assumption of "progress." That is, so-called "higher forms" are taken as improvements over "lower" ones. This metaphorical idea of a ladder of development is still prevalent today. Studies of how different publics view applications of genomics show that while manipulations of "lower forms" are acceptable, our discomfort increases as genetic changes are conducted between animal species or genes are moved from humans to non-human animals or vice versa. This hierarchy of acceptability may be influenced by religious beliefs or cultural ideas and preferences.

Findings of genomic science, however, are contributing to dislocations of scientific and public imaginations and understanding. The original speculation that humans had close to 100,000 genes has since been displaced by a humbler picture of the human species: we are only a complement of about 30,000 genes, a figure that surprisingly is similar to the number of genes in the rat or the mouse. Scientists estimate that our genes may be 70 to 90 per cent similar to mice and our genes also share a 95 to 98 per cent similarity with apes (chimpanzees, gorillas and orangutans). Subtle changes that have accumulated in the genes of *Homo sapiens* and these other animals have led to quite different organisms. By looking at how genes are expressed—that is, the activities involved in genes sending directions to cells to make proteins and related compounds—we are gaining a better understanding of how groups of organisms share similarities or differ.

The "tree of life" is another metaphor that has been used to capture our understanding of evolutionary change. In an illustration of Darwin's *Origin of the Species*, the picture of a tree of life is used to show a pattern of diversification, with the tree's branches increasingly diverging from the trunk and never connecting. However, our understanding from evolutionary biology and genomics is showing us that this tree of life metaphor does not adequately capture the richness and complexity of evolutionary patterns and processes.

We are learning that the evolution of species, particularly among bacteria and other micro-organisms, is more characteristic of a web that crosses and re-crosses through genetic exchange, even as it grows outward from a point of origin. While we have always known that genes are passed on "vertically," that is, from parents to subsequent offspring, we have more recently learned that genetic exchanges can also occur between species. In microorganisms, this phenomenon of horizontal exchange is the explanation given for why antibiotic drug resistance has developed in bacteria or why new pathogens emerge. Scientists suggest this may be how we humans have also developed our acquired immune system. We may have pieces of genetic material from our "lowly" cousins within us, which challenges our notions about the distinction of species. These genetic reorganizations have occurred in large part because of changes in environments.

The web of life metaphor has also helped to illustrate our understanding of biodiversity, with genetic diversity as one of its strands. Changes in the genes of organisms that result from mutations, gene exchanges, and dynamics are part of the picture of genetic diversity, but these changes are complemented by the anchoring strand of ecological diversity, the interaction between species and their environments. It is this dynamic interaction between the genetic and the environmental pictures that makes the web of life a complementary metaphor. Such interaction is also a reminder that our place in this web is very much linked to and interdependent on our environments and the rest of this intricate web.

Metaphors are ways of seeing; they are also guides for action. We have seen ourselves as the fittest of survivors at the top of an evolutionary ladder—but the lessons we are learning at the gene's eye level challenge this picture. We don't even hold a candle to our bacterial cousins in the game of adaptability! The web is a new signpost in this metaphorical journey—but it requires shaking off our hubris and rethinking our place at "the top."

JIM EVANS

THE DARK AGES were aptly named, for they were dark in both a literal and a metaphorical sense. Before the advent of science, droves of children died of pertussis, appendicitis was a lethal disease, and when we left our families to move a few hundred miles away, the chances were good that we would never again hear their voices, let alone see them again. In addition to being in the dark literally, we were also in the dark figuratively; we did not understand the simplest workings of the world. From Newton to Darwin, science has enriched our understanding of the universe as it has also made our lives materially more comfortable. And while science has brought its share of tragedy and heartache, I doubt that many of us would trade our life for that of even the most privileged citizen of the thirteenth century. Philosophically, too, the successes of science have had profound implications. The fact that with the touch of my foot I can propel my car to eighty-eight feet per second tells us something fundamental about the world in which we live and about science itself: our ability to light up a room, cure disease or fly demonstrates the existence of a remarkably consistent external reality. Science is a portal, a keyhole, through which we have learned to glimpse that reality.

Science is an intellectual "tool kit" that humans have developed for the purpose of explaining and controlling our environment. It is an intensely practical endeavour and one that rests firmly upon a set of rules and procedures. This intellectual toolkit consists of observation, hypothesis, measurement, prediction and, perhaps its most critical component, experiment. Critically, conclusions in science are always tentative; there is no place for dogma. This tentativeness, coupled with its cumulative nature, gives rise to one of science's most important attributes: it is self-correcting. Newton's laws of motion are useful; Einstein's are better. Mendel's explanation of genetics was profound; modern genetics more precisely describes heredity. The stuttering course of science gradually but inexorably inches us toward a more accurate view of the universe. This same principle of tentativeness also rules out invocation of the supernatural or personal revelation. Note that science does not say there is no supernatural and that people can't have revelations. What it does say is that such things are not science.

Science differs from other human pursuits in that it is not an arbitrary belief system or a purely social construction. Take, for example, another human endeavour that so enriches our lives, art. Whether I am moved by Jackson Pollock or my tastes run to Vermeer is a purely subjective matter and is not tethered to an objective underlying reality. But to put myself in an airliner and hurtle at 600 mph across a continent 30,000 feet above the earth's surface means that I acknowledge (with my very life) that science is able to glimpse and control the consistent and underlying reality of the universe.

Like science, art is an exploration of our world. Both are intensely creative pursuits and both are beautiful in surprisingly overlapping ways. The beauty, for example, of the DNA double helix is a vivid reminder of that overlap. However, science is universal while art is uniquely human. Here lies the critical difference between the two, and contained within that difference lie the strengths and limitations of each. The melting point of cesium and the speed of light appear to be the same here as in the Andromeda galaxy. Science is a process by which we explore the underlying reality of the world; its discoveries are independent of the human mind. While this gives science great power, it also is true that science cares not for humanity. In contrast, art is the quintessential human endeavour; an exploration not so much of our world but of ourselves. Art is a process by which we explore our unique (and, critically, our common) perception of the universe. Its intent, as E. M. Forster admonishes us in the epigraph to *Howards End,* is to connect us both with our world and with our fellow humans.

At its essence, the purpose of art is to invest our lives with meaning. While strictly practical criteria define what is and what is not science, art is not shackled by such rigid criteria; it is a pure product of the human mind and culture. Its only rules are that it must evoke emotion and resonance. The universe simply "is" and science is our way of knowing it. But the universe of art is infinite, defined and limited only by the human mind. *We* define artistic reality, as Duchamp so elegantly demonstrated with his urinal cum art.

If, one day, we finally stumble upon differently evolved beings elsewhere in our galaxy, the idea that a Hopper painting or a Beethoven sonata will deeply touch them is as unlikely as the proposition that their fundamental laws of motion will differ from ours. Our science, but not our art, is likely to be mutually intelligible, for science is universal and glimpses the common underlying reality of the universe, while art is a pure product of the human mind. But I'm guessing that they, too, will have developed their own art. For while beautiful, awe-inspiring and creative, in the end science alone is not enough. Science can demonstrate our need for emotional warmth and connection. It can even explain (through neurobiology and evolutionary biology) *why* we need these things. But it cannot satisfy those needs. For meaning and fulfillment we need one another and we need art.

Royden Mills
No Longer Between
(Sculpture in foreground of
studio installation view), 2008
Mixed media
Courtesy of the artist
Photo: Mark Freeman

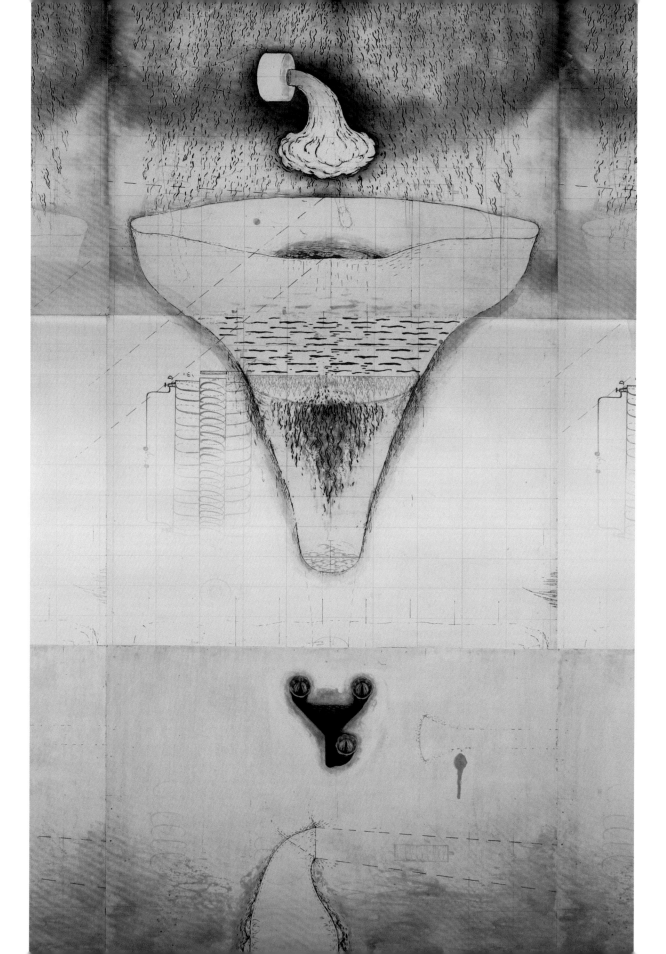

Sean Caulfield

Body Landscape Diagram

(Lethe Suite), 2007

Mixed media on paper

Mounted to canvas

107 x 166 cm (42 x 65.25 in)

Courtesy of The Scott Gallery

Photo: Mark Freeman

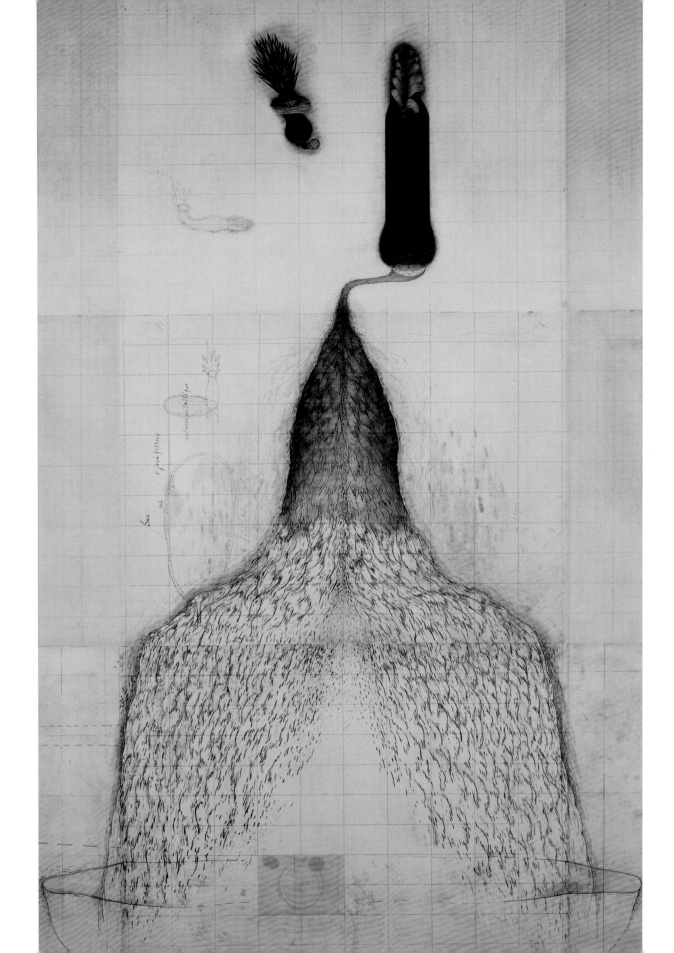

Sean Caulfield
Descent Diagram
(Lethe Suite), 2008
Mixed media on paper
Mounted to canvas
107 x 166 cm (42 x 65.25 in)
Courtesy of The Scott Gallery
Photo: Mark Freeman

Sean Caulfield and Royden Mills have worked collab-
oratively to produce a series of two-dimensional and
three-dimensional works that are loosely based on images
from the history of scientific and medical illustrations.
Their drawings and sculptures suspend viewers between
a number of associations, including references to historic
scientific objects, fictional science, biological forms, as
well as bio/mechanical dichotomies. They are interested in
examining the ways in which historic scientific images and
objects, intended to convey an objective truth about the
world, can in a contemporary context become more appreci-
ated for their aesthetic/poetic value.

DAVID GARNEAU

ART AND SCIENCE are mutually exclusive fields. So-called art/science collaborations are inequitable: one discipline always dominates by using the other as a tool. Such "collaborations" are more expressive of a desire than a logical possibility. What many artists desire from the exchange is to insert themselves into the debates that arise from scientific inquiry and the application of scientific results. Artists do not really engage science so much as the ethics of science, and while art does not have much effect on science, ethics permeates and influences the agents of both fields.

Science is a system that uses observation and experimentation to describe and explain the material world. It continuously corrects and improves; it aspires to objectivity. Science is also the literature produced by people using the scientific method. Scientists are required to speak the same language (mathematics, for example), know its histories (the relevant literatures), its internal disciplinary borders, and maintain its external boundaries (metaphysics, emotions, biography, art, and so on).

Art has no agreed upon definition, no common system, methods, goals or boundaries. It admits the possibility of nearly anything. There are literatures *about* art—art history, criticism, philosophy of aesthetics—but they are not art; they are meta-discursive disciplines that have art as their subject. Art is subjective, expressive, usually imitative, often

fictional, unsystematic, unconscious and extrarational. Art is not a language; therefore, artworks are not propositional. They may inspire, illustrate and communicate knowledge, but they do not produce it.

Science rarely crosses into the art realm—except, perhaps, to explain how the style of Monet's late landscapes are due to cataracts, how Van Gogh's *Starry Night* can be accorded to migraines, and El Greco's elongated figures to astigmatism. Individual scientists occasionally use scientific imaging tools to produce stunningly beautiful images that they describe and even display as art. While definitions of art are elusive, art institutions are more conservative and rarely embrace such works. Curators and aesthetic philosophers argue that art has not been synonymous with beauty for a very long time. Some artworks are beautiful, but not all beautiful things are works of art. This may be a current prejudice. Because art is fluid, any thing, arguably and eventually, *could* be art. Even so, the inclusion of the pictures in this book would not mean that science is art, perhaps only that some scientists are now considered artists. The images are not science, only the artistic result of using scientific tools: art can be made by any means and materials. The divide remains.

Scientific illustration might look like art. Drawing is an art form, but not all drawings are art. Illustration is its own discipline. It is a technology or craft between, in this case, art and science. Illustration can be employed as a descriptive tool of science. Too much art (creative interpretation and expression) weakens an illustration as a tool for science. Too much accuracy, no metaphor, personality or subjective play, makes for a poor work of art.

Science is generally conscious of its limits. Art, however, rarely recognizes boundaries. Most art, as Plato complained, is imitative. It pretends to be other things all the time, including science—but looking *like* science does not make it science. Just as beauty is often mistaken for art, technology is often confused with science. Many contemporary artists employ technology: GPS systems, radiological imaging, computer engineering and so on. These artists are not doing science; they are just borrowing its tools.

While artists may describe their work with scientists and engineers as collaborations, in fact, they are using their craft as a tool for science, or using scientific tools for artistic means. In the first case, scientific knowledge may increase, but there is no contribution to art. In the second case, art may be advanced, but science remains unperturbed. A true art/science collaboration requires both systems to be affected and, hopefully, advanced.

Given that art and science are antithetical, their meeting ground must be on a third field that influences both, for example, ethics. Through plastic surgery and psychoanalysis, the performance artist Orlan is transforming herself so thoroughly that she hopes eventually to warrant a new legal identity. Her work is powerful because, if we can face it at all, we are forced to consider the ethics of her self-

abuse and its reverberance for the whole realm of elective surgery. She elicits limits by crossing them. While Orlan is not doing science, and might not even be doing art (but who could say she is not?), she *is* performing an aesthetic ethics that bridges and could affect both art and science.

Leonardo da Vinci's wonderful anatomical drawings seem to be the great exception that combines art and science. In that moment of recording his observations, he seems both an artist and scientist. However, again, he is really, at that moment, a scientist using his great drafting skills to record scientific research. When we recognize these objects as art—there is no evidence that he thought as much—we are not seeing them as scientific research. The scientific gaze and the artistic gaze use the same works differently. Science is looking for material fidelity; art is looking for expression and metaphorical meaning. Read as an aesthetic ethics, the drawings have us wonder about the propriety of a man cutting open the pregnant belly of an unconsenting woman to compare her dead child to the fetus of a horse. Such works are performative of an anxious presence that demands the intervention of an ethical consciousness, which is provided, not by the artist, but by the viewer.

Ethics asks the scientist and artist equally to consider the implications of their research within the large human field. Because science is a true discipline, codes of ethical conduct are conceivable and enforceable. Contemporary society is more reluctant to control art. Because art is metaphorical and non-propositional, it persuades by evoking deep thinking and feeling in the viewer. Andres Serrano's autopsy photographs do not tell us how to think or behave. They shock us into developing our own thoughts and behaviours. Many artists resist ethical guidance, and so some should, but if they conceive of themselves as doing aesthetic ethics they might guide themselves. With a firm ethical sense, they might produce artwork that could impact scientists and science, as well as art.

GAIL GELLER

"There is no use trying," said Alice, "one can't believe impossible things."

"I dare say you haven't had much practice," said the Queen. "When I was your age,...sometimes I've believed as many as six impossible things before breakfast."

—LEWIS CARROLL, *Alice in Wonderland*

HOW MANY OF YOU have practised believing something impossible before breakfast? In embracing this quote, I am not advocating irrationality or rejecting the beauty, elegance and importance of scientific "truths." Rather, I am arguing that we cultivate a higher tolerance for uncertainty and curiosity about the mysteries in life. Uncertainty is ubiquitous in biomedicine, generally, and in genetics in particular. Decision theorists distinguish two components of uncertainty—risk and ambiguity. Although both are associated with circumstances in which there is more than one possible outcome, risk exists when the probability of each outcome is known. Ambiguity exists when the probability of each outcome is unknown. Both are relevant to the communication of probabilistic information associated with genetic testing for susceptibility to complex diseases and traits. For example, an individual whose parent carries a cancer-predisposing mutation has a 50 per cent risk (known probability) of inheriting the mutation. By contrast, a woman who carries a mutation on one of the breast cancer genes may be told that she has anywhere from a 50 to 85 per cent chance (an ambiguous range) of developing breast cancer in her lifetime. There is no guarantee that genetic testing will identify an inherited susceptibility mutation, if one exists, or that carrying an inherited susceptibility mutation will result in the development of the disease. Therefore, those who are involved in decisions about genetic testing for cancer susceptibility have to grapple with various uncertainties.

My empirical research in bioethics has focused on communication and decision-making under conditions of ambiguity or uncertainty in the clinical and research contexts. I am fascinated by how people perceive and respond to uncertainty. With specific reference to genetics, I am curious about the impact of uncertainty on the informed consent process for genetic testing, on media coverage of new genetic technologies and on the well-being of genetics professionals themselves. My theory— a "testable hypothesis"—is that clinicians abhor uncertainty and researchers embrace it.

Sociologist Renee Fox and psychiatrist Jay Katz offer important insights regarding clinician reluctance to acknowledge the uncertainties in medicine. Both argue that physicians are trained, or socialized, to deny uncertainty—indeed to substitute certainty for uncertainty. Physicians are afraid that if they allow themselves to entertain questions and doubts, they will no longer act with conviction, and their effectiveness will be diminished. Professing certainty allows physicians to maintain both an "aura of infallibility" and professional power and control over the decision-making process. Clinical judgement—the "art" of medicine—is rarely acknowledged to patients.

Unlike most practitioners of clinical medicine, scientists are self-selected because of their curious nature and are trained to value the questions, as well as the answers—to think "outside the box." Every major advance in scientific thinking—so-called scientific revolutions—has resulted because someone dared to believe the impossible. Consider the transformation from Newtonian physics to Quantum physics—from a focus on matter to a focus on energy. Quantum physics has discovered that matter/energy does not exist with any certainty in definite places but rather shows its "tendencies" to exist (the Uncertainty Principle). Moreover, the universe—or the existence of matter and energy—is fundamentally dependent on the existence of an observer (the Observer Effect). In other words, what we observe is what is "true." The notion of universal truth is being questioned.

We are on the brink of another scientific revolution. The medical and scientific establishments herald an era of genetic medicine, when an individual's genetic sequence, or genotype, will inform his or her diagnosis, prognosis, treatment options and behaviours. In order for this revolution to take root, two obstacles will have to be overcome. First, we will have to let go of our dogmatic reliance on the randomized clinical trial (RCT) as the gold standard of "evidence-based medicine" and explore creative modifications to our scientific methods. The RCT is founded on several fundamental assumptions, such as homogeneity of the study population and standardization of the intervention, that are inconsistent with the goals of personalized medicine. Personalized medicine is based on the premise that genotypes are unique and interventions should be tailored, much in the same way ancient systems of healing have always personalized treatment based on the particular manifestation of disease in each unique individual. Ironically, conventional medicine has rejected the person-specific approach of unconventional treatments and rendered the individuality of response to them as "unscientific" while simultaneously embracing the person-specific goals of genetic medicine.

The irrational reluctance to incorporate interventions that may be useful, such as those used in traditional forms of healing and the blind willingness to incorporate scientific discoveries (genetic testing) that may not be useful are motivated, in part, by fear of what we cannot explain or understand. Such fear not only fuels the debate about the scientific validity of ancient systems of healing but also animates media hype about new biomedical discoveries. Overcoming our collective fear of the unknown is a second, perhaps more intractable, challenge to embracing revolutionary scientific thought. Art can help; it has the potential to transform our fear of uncertainty into an appreciation of mystery by offering us the opportunity to experience wonder and awe, and to "practice" believing impossible things.

"The most beautiful thing we can experience is the mysterious. It is the source of all true art and all science…"

—ALBERT EINSTEIN

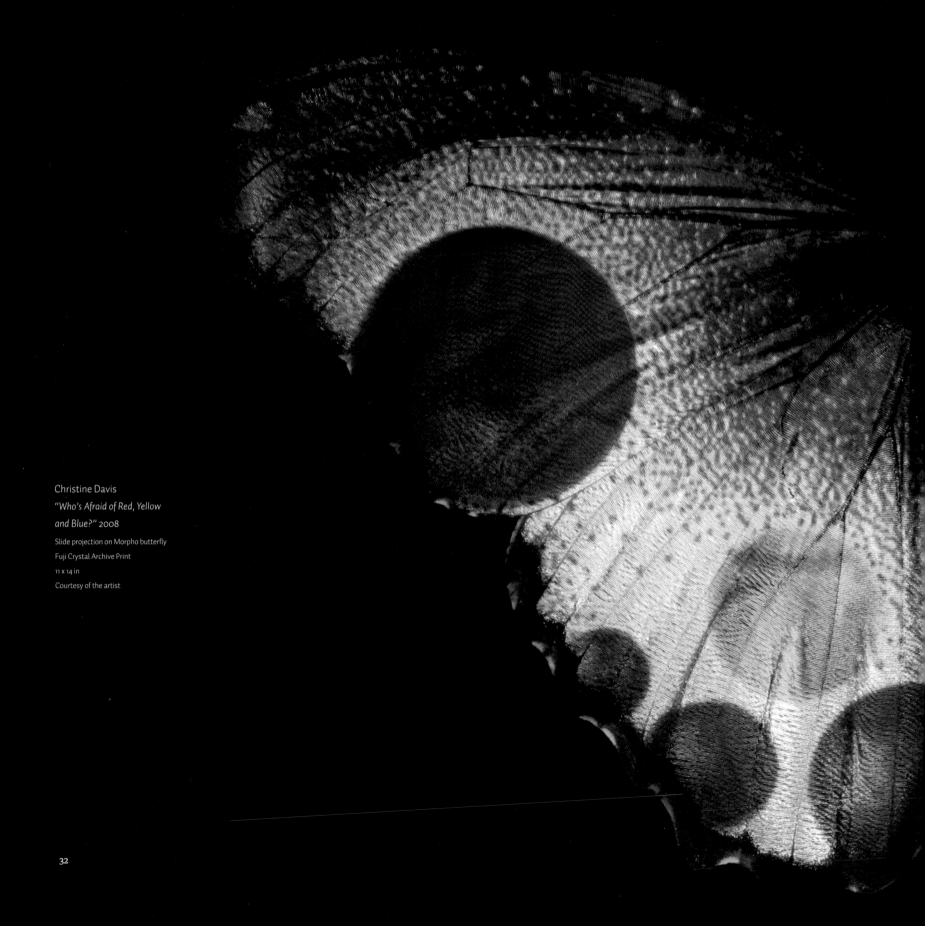

Christine Davis

*"Who's Afraid of Red, Yellow
and Blue?"* 2008

Slide projection on Morpho butterfly

Fuji Crystal Archive Print

11 x 14 in

Courtesy of the artist

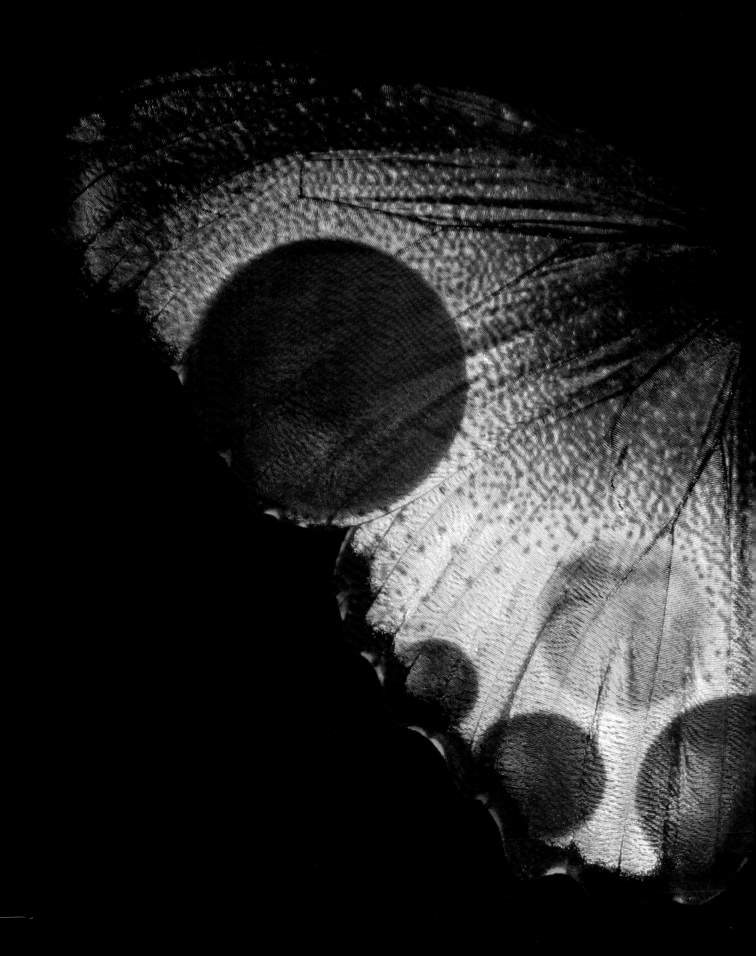

Christine Davis's *Who's Afraid of Red Yellow and Blue?* is a slide dissolve projection of still photographs onto a single, bright blue, Morpho butterfly, which makes use of "low" tech analog projection and digital controls to create a fragile, haunting image that is in a continual state of transformation. The image evokes a world in which dichotomies coexist, a place where animate/inanimate, material/immaterial and the micro/macro are in a state of constant flux and transformation. The viewer is given a contemplative space in which to consider how our own intellectual projections inform our experience of the world. Stated differently, in Davis's own words, "Within my practice the fragility, finality and material dependence of the projected light/image extends metaphorically and processorially to the life world. I approach the projection screens as living assemblages; processing different fragments, orders and experiential modes of existence. Watching one is like looping between perception, hallucination and cognition. A vertiginous gap opens between feeling and knowing. And the loop keeps going, outlining the question: What is human?"

HANK GREELY

"There are more things in heaven and earth, Horatio, than are dreamt of in your philosophy."

—WILLIAM SHAKESPEARE, *Hamlet*

THE WORLD OF BIOLOGY is more complicated, and more bizarre, than we imagine. Public reactions to bioscience are often based on oversimplified biology, where both organisms and species are separate and well-defined, and all life basically acts like humans. This view underlies the "yuck" factor: the public's visceral reaction against some biotechnologies. But biology is not what it seems, and neither are we. Art might help us understand that.

I am made up of about 100 trillion human cells. Some of those cells are helping me write these words by thinking them and moving my fingers on the keyboard. Others provide essential support, keeping my body functioning. Still others are doing nothing. Some may even be undermining me by malfunctioning—not fatally, I hope.

Every hour I'm alive, millions of my cells die, but when I die, some of my cells will live on, for minutes, hours, even days. If a cell line has been made from my cells, they may live indefinitely. Almost all of those cells contain all of my genes. Am I the organism, or the cells, or both? Or am I, possibly, like my cells, part of some larger organism that lives or dies apart from me?

Of course, not all of my human cells may be mine. If I had an organ transplant—or even a blood transfusion—another person's cells are working inside me. Even without transplants, some non-identical twins have cells from their twin sibling thriving in their bodies. Those twins exchanged cells *in utero*, becoming mixtures, usually slight, of each other. Some people without twins show the same pattern, probably as the result of a twin who died early in pregnancy. And mothers sometimes carry cells from their children for decades as a result of fetal cells crossing the placental barrier into the mother's blood.

Such people are intraspecific chimeras, individuals made up of cells from more than one member of their species. Who are they?

But we don't need to stay within our species. Hundreds of thousands of people are walking around with, and because of, pig valves in their hearts, valves that are effective replacements for human heart valves. Are those people, at heart, part pig? They are, most certainly, interspecific chimeras: creatures made up of cells from two different species.

More broadly, our human genome contains few, if any, "human" genes. It contains human *versions* of genes found in monkeys, mice, flies, yeast and even bacteria. All living things are, quite literally, cousins; sometimes we humans share large stretches of genetic sequence that are identical to those of cousins both near (chimpanzees, with whom we are more than 98 per cent genetically identical) and far (fruit flies, whose homeobox gene is almost identical to one of ours). These genetic similarities are ancient genealogy—we truly *are* cousins—but others are more recent, genes smuggled in by viruses and incorporated into our genomes. Sometimes, we do this intentionally as gene therapy. More commonly, it's an accidental result of a viral infection. And if the virus gets into an egg or a sperm cell, this new viral gene can be passed down through human generations. There are relics of old viruses in our "human" genomes, some possibly from our lifetimes, others from a distant past. Genomically, we are all interspecific chimeras.

Perhaps we go wrong by thinking of ourselves as organisms at all. I have about 100 trillion human cells, but I have more than 100 trillion non-human cells, mainly bacteria and much smaller than my human cells. They make up about one per cent of my mass, living, eating and reproducing happily in my intestines or my mouth and in other friendly nooks and crannies of "my body." I may think they are interlopers; they would think I am an ecosystem.

These non-human cells are not just bystanders. Some help me digest food; others occasionally make me sick. They might even have deeper effects. One protozoan, called *Toxoplasmosis gondii*, lives primarily in cats. In order to reproduce, it needs to leave a cat and spend some time in a rat. It does this in cat feces, which sometimes infect rats. After a few days in the rat, the protozoan is ready to reproduce, but it needs to get back into a cat. So the microbe makes the infected rat lose its fear of cats—and end up as cat food. About one third of all humans are also infected with *Toxoplasmosis gondii*, which usually seems to have almost no physical effect on us. Does it change *our* behaviour? We don't know. Yet. The invisible life within us could affect us, as well as

infect us. Where I end and other life begins is not clear.

And so it goes throughout biology. Sex between a life-long male and a life-long female is not the only way life reproduces. Even in humans about one per cent of babies born in developed countries are the result of *in vitro* fertilization, not sex.

Some other species never use sex for reproduction, others switch between sex and cloning, and some always have sex but only with themselves. For some animals, sex is determined after birth by the environment—and in some of those it changes several times during the individual's life. This all happens in animals we can see. Plants, fungi and bacteria can be even stranger, sometimes with human help. The red wine that is helping (some of) my cells complete this essay comes from the branches of a French Pinot Noir grape vine that were grafted onto roots and branches of a North American vine of a different species and genus. I drink the product of a cloned, chimeric plant—and I like it.

"Pardon him, Theodotus; he is a barbarian, and thinks that the customs of his tribe and island are the laws of nature."

—GEORGE BERNARD SHAW,
Caesar and Cleopatra

The biological world is a strange place. Even our tiny corner of it can be very different from what we understand; the farther shores are truly weird and wonderful. Products of biotechnology may be good or bad, helpful or dangerous. They should be judged on that basis, not because the methods are strange and unsettling. That they seem "yucky" to us may be our own fault. The ability of art to illustrate the invisible, to demonstrate the unforeseen and to juxtapose the incompatible may help us understand these broader biological realities and keep us from rejecting useful innovations because of our own provincial ignorance.

JONATHAN LOCKE HART

1.

In life we never know much beyond the surface of things
In quest of the nature of things, who speaks and who sings,

Who knows and who does not, the way the prepared
Find luck because they have worked, fought and dared

To push at the known world. The mind is as bold
As the hand: the curious never get old.

2.

Aristotle sought to investigate life,
Physics, poetics, and before the strife

Tutored Alexander who changed war
And politics, who marched and died before

He could take India. Empires come and go
Napoleon's soldiers died in the snow.

3.

Archimedes explained the lever, advanced
Calculus, and for his machines was lanced

By a Roman soldier against orders. So much
For obedience, a level, a crutch

To upset equilibrium. Fluid statics
He studied pressure before dynamics.

4.

Forces are balanced in physics as in war
If only the human world might take store

In being equal in all directions
An isotropic wish for our sins.

Galileo, who worked in this field, knew
How hard contesting worlds battle and sue.

5.

Pascal extended Archimedes' work
And few could ignore, evade or shirk

His brilliance. What came from Syracuse.
Was hard to deny. Why accuse

Science of defying nature
Denying the divine? What a stir

His death ray made. Was the burning glass
A myth like leaves of grass

As Descartes thought? Had the mirror burned
The Roman ships as they tossed and turned?

6.

Aristotle brought method to the study of life
And Owen said zoology came not like a knife

But sprung from this Greek like Athena
From the head of Zeus: I translate what I saw

As a story of how this Attic sage could look
Behind seeing and the law of the book

And find organized knowledge. He sought
Beyond the received, what was and what ought

And turned from Plato's paradox of inquiry
And formed a philosophy of biology.

7.

So in Aristotle inquiry became the key,
Historia, which also meant history

And story. And so the myth of science grew
From a common root, and all that would ensue

Would mingle between story and proof
And rather than remain on high or aloof

Aristotle worked hard on form, explanation
And necessity, until his practice was done.

8.

Admirable Aristotle, master
Of many fields, could distinguish and infer

So much from what he observed, science and the arts
All one. His systematics, the whole and its parts,

Brought light to animals, poems and polities
And taught us about ethics and disease.

9.

Blake chided Locke, Newton and Voltaire
And romantic art split from science in a dare

But how would we do without understanding
How the laws of motion then mechanics could spring

From physics and mathematics? Is this not
Gravity enough? Why do some grow so hot

Before the power of science? Nothing is good
Or bad but thinking makes it so. The wood

Is not dark because we cannot see
Beyond the horizon to the winedark sea?

10.

Rhyme has no reason but does. Newton took
To time in physics and theology. His book

Was open to word and world, visible
And not. So much is beyond, invisible.

We read Newton's originals in school
Prisms and telescopes, exceptions to the rule.

11.

Pingala, Omar Khayam, Yang Hui
Expanded the power of sums: Moriarty

Wrote a fictional treatise on this theorem
And Holmes used logic to solve a conundrum

And is loved in the threshold between
Words and numbers, the seen and unseen.

12.

Which brings us to Lyell and Darwin
In some roundabout way: my sin

Is being a poet who liked numbers and physics
And tried not to get too lost in metaphysics.

Leaving Kant aside, taking Shakespeare up,
Many things fall between lip and cup:

We wander between earth and heaven
Born without choice, seizing choice. Even

Hermes was given to half-rhymes. Pope
Alexander and Alexander Pope

Divided texts that divided worlds. Epic catalogues
Burst the history of biology with epilogues

Based on Vesalius, Harvey, Linnaeus,
Lavosier, Watson and Crick, the fuss

Over DNA, cloning and engineering
What makes some cry and others sing.

13.

Mendel and Morgan did the early research
And controversies threatened to besmirch

The good work done. Eugenics, racism,
Biological warfare created a schism

By obscuring knowledge with ignorance,
Wisdom with ideology. And since

The ethics of cloning and gene therapy
Has come under increasing scrutiny.

14.

What would Marie Curie think of Hiroshima?
Einstein later regretted, if Pauling's words are law,

Signing the letter to Roosevelt. We are all
As conflicted as Hamlet: death and the call

To life in love and strife is no simple thing
To cry in whispers or in proclamation sing.

15.

I have sat by those who first found clone
And pulsar, fine people of flesh and bone.

Let us make this human, as if in school or home
We had friends and family suffering, some

Getting better, others not, who needed help
The past could not give but some great leap.

16.

What are we to do when we look into the eyes
Of those we love and wish to make them well? Skies

And lungs will not clear unless something is done.
We have read Frankenstein, *and know the sun*

Is a fire that Prometheus carried home. The quest
Of Helios we seek from east to west.

17.

Evil is an awful word. History
Is full of waste and abuse. Listen and see

How war has been made of science, but so much peace
And good have come of it. How do we release

That good, constrain the tyrant within us
And without amid the clamour and the fuss?

18.

In this human story as it unfolds, some say comedy,
Some tragedy, some divine, others a remedy

Through the beauty and elegance of nature,
Can we use knowledge as a suture

Not abuse its power? Can we revive
And find a way to prosper and survive?

19.
We have spirits to feed. We have people to love.
Science is not ignorance's scapegoat. Above,

Behind, below, ahead, what awaits
As they leave a myth beyond the eastern gates,

The tree of the knowledge of good and evil
At their backs, determined or with free will?

20.
The dilemma of science comes down to here and now
And we need to imagine it and not make it a why and how

We are blind and deaf to. We all feel strongly
But let us not leave reason, judge wrongly

And weigh what we can do in good conscience
A science together with no easy end.

December 9, 2007

Lyndal Osborne
ab ovo, 2008
Mixed media installation: glass, foam, and paint
20 x 10 ft (detail of a work in progress)
Photo: Mark Freeman

47

Lyndal Osborne
Endless Forms Most Beautiful,
2006
Mixed media installation
7 x 25 x 15 ft
Photo: Hide Away Studios, Medicine Hat

Lyndal Osborne has developed an individual approach to
her sculptures and installations that utilizes found and
recycled material, which she alters through the application
of colour, manipulating their original shape, and/or by
placing the objects in new contexts so that they develop
new metaphorical meaning. This evocative installation work
speaks poetically of the forces of transformation within
nature and comments on pressing issues relating to environ-
mentalism. In her recent work, Osborne has focussed on an
examination of genetically modified organisms as a source
for subject matter.

ART AND SCIENCE

Conflict or Congruence?

ANNA R. HAYDEN & MICHAEL R. HAYDEN

ART AND SCIENCE are commonly contrasted by their differences, with science representing the product of the rational mind and art the expression of the creative spirit. As such, these two fields of endeavour are frequently portrayed as occupying two solitudes with little communication between them and apparent conflict being present, as they compete for the share of the public's heart, mind and commitment.

We have, however, been struck by what is shared between art and science, both at the core representing products of the human spirit. Both art and science represent the pursuit for truth, changing the way in which we view ourselves, and bringing new understanding to the world at large. These two disciplines are ultimate expressions of individuality and imagination, reinforcing the value of each person, and reminding us of the profound nature of all creative activity. As such, these activities contribute to the much needed rebirth of the human spirit. Both art and science manifest as expressions of intellectual and emotional experience, and both benefit from expanded knowledge and breadth of experience, often in fields different to the central area of concentration.

Pivotal art and science always requires courage as it confronts the established way of thinking and forces us to refashion our way of relating to the world. The most profound art and science is often first met with considerable resistance, even direct opposition, which may seek ways to undermine the artist/scientist; such was the case with Van Gogh, who was completely ostracized in his time, or with Galileo, who was put under house arrest during the Inquisition for daring to suggest that the planets moved around the sun and not the earth. In many ways, revolutionary concepts are ahead of their time but their endurance and influence ultimately attest to their originality and truth. For the individual artist or scientist, the early part of the conceptual journey is often painful and lonely, but the pursuit, independent of financial gain or reward, is necessary for their creative fulfillment.

While intuition is often recognized as integral to the arts, intuition is frequently also crucial in science. With so many unanswered questions and different methodological approaches to be pursued, intuition often allows the scientist to choose the path that can and must be answered. The challenge for the scientist is to remain deeply open to that intuitive spirit, to allow it to act as a guide at important forks in the road. Exposure and immersion in the arts helps to keep that spirit alive and present. In this way the arts are integral to and enrich successful scientific exploration.

In history, times of great artistic endeavour have often emerged during periods of profound scientific ideas, with considerable interchange between them. Examples abound including Ancient Greece, the Renaissance, Mayan culture and periods in Chinese and Indian history. It is therefore clear that what we require today are more opportunities for significant interaction between artists and scientists with the power to enrich each other's creative lives.

Science/art interaction is not merely achieved by having works of art present and displayed in places of scientific exploration. Rather, this may be accomplished through establishment of new forums for direct and open communication between artists and scientists, sharing perspectives and exposing each other to their creative works resulting in joint projects and collaborations.

What a thrill and how illuminating it has been to hear an artist's expression and reflection of what is observed in the science laboratory, such as perspectives on what view is seen while looking down the microscope. Such insights can profoundly alter what is perceived for both artist and scientist and can lead to new forms of interaction in unexpected ways.

One reflection and recognition of the need for more interaction is the development of undergraduate arts and sciences degrees, which are still in their infancy. In most instances today, courses offered are taken from those routinely provided in the arts or science faculties. But what is needed are combined courses taught by individuals involved in this integrative path; they have a unique perspective and are crucial to the prosperity of such important programs. Practitioners of the arts and sciences are needed to combine these disciplines and to challenge students to be at the vanguard of this essential but nascent activity. We must ensure that students of talent and ability do not have to make difficult choices between the arts or sciences but rather are able to experience and learn that which is already clear to a growing number, that both are essential for successful pursuit in each individual field.

Society today needs to be reminded that the individual is of greatest value. We argue that this approach is essential, needs to be more widely embraced, and can contribute profoundly to the human spirit and culture of our society. Societies can be judged in part by their support of the arts and sciences. This is reflected by the place of these ideas in the prevailing government and by the levels of direct support given to the arts and sciences. This cannot be a luxury or be given only in times of prosperity, when resources are plentiful. Rather this is an important sign of the health and well-being of the nation. The times of greatest achievement in history are paralleled by creative output in the arts and sciences. We would do well to remember the remarkable Gabrielle Roy quote on the Canadian twenty-dollar note: "Could we ever know each other in the slightest without the arts?"

The challenge posed recognizes the need for integration, and it emphasizes the place of the arts in all of our creative endeavours, particularly the sciences.

JAY INGRAM

IN THE EARLY 1970S, Gunther Stent, one of the instigators of the molecular biology revolution of decades before, argued that there is an unacknowledged congruence between creativity in the arts and science. Stent pointed out that it is widely accepted that if Shakespeare hadn't written *Macbeth*, no one would have. But, he asserted, the opposite view held of science—that if Watson and Crick hadn't discovered the structure of DNA, some one else would have—was dead wrong.

In his view, things would have been significantly different if Watson and Crick hadn't published their page-and-a-half in *Nature* describing, for the first time, the now-famous double helix (accompanied by a precision drawing by Crick's wife, Odile). But Linus Pauling, Rosalind Franklin and Maurice Wilkins were all hot on the trail of DNA, and while no doubt Pauling had the intellect to figure it out, the Brits had the data. So what difference would it have made if someone other than Watson and Crick had "discovered" DNA?

According to Stent, plenty. He first dissected what it means to say that any of Shakespeare's plays are, in fact, unique. Not from the point of view of the plot: Shakespeare reworked earlier plays going all the way back to the ancient Greeks. Not even from the point of view of his choice of a human flaw and the emotional wreckage it creates. Those, too, were not new. Stent eventually reduces it, at least

in his mind, to the fact that no one else could have combined poetry, language, drama and emotion as beautifully as Shakespeare did.

But you could exalt Watson and Crick's powers of expression, too. Stent says that no one else would have come up with not only the structure of DNA but its implications, in the pair's famous phrase, "It has not escaped our notice that the specific pairing immediately suggests a possible copying mechanism for the genetic material." A series of different investigators might have only peeled back the onion layer by layer, while others would have been muddying the waters at the same time. Stent quotes Sir Peter Medawar with approval: "The great thing about [Watson and Crick's] discovery was its completeness, its air of finality...If the solution had come out piecemeal instead of in a blaze of understanding, then it would still have been a great episode in biological history; but something more in the common run of things; something splendidly well done, but not in the grand romantic manner."

The grand romantic manner? Hardly a phrase that is commonly applied to science. The irony here is that it became clear to Stent that few scientists actually shared his opinion, reverting instead to the age-old view of which Stent was so tired—that if Watson and Crick hadn't figured out DNA, of course others would have, and science would have gone on unimpeded.

Stent, in trying to understand why scientists were out of step with him, introduced a couple of intriguing ideas. One is that while science is obviously cumulative, most scientists don't realize that art is, as well. Today no undergraduate biochemistry student has to read Watson and Crick's paper because the revelations in that paper have been repeated and absorbed by the scientific community. But Stent argues that art is the same: while it's true that Beethoven didn't supersede Mozart, he and all other composers have built on the music written before them.

He also points out that art and science have different susceptibilities to *paraphrase*. The essence of a scientific act of creativity can, as time passes and familiarity grows, be reduced to a single statement: "DNA is a double-stranded self-complimentary helix." But a work of art cannot be so rendered. To rewrite a Shakespearian play would take a genius of Shakespeare's stature.

I quote these ideas of Stent's because, regardless of whether you agree with him about the uniqueness of scientific discovery, it's refreshing to hear that science is not a mechanical, inevitable process of turning rocks over and discovering what lies underneath. It is a creative and imaginative world of ideas; in Watson and Crick's case, as pointed out by Matt Ridley, the creative act was performed in the heat of an intense race to be first.

Like the arts, science is also a forum where the expression of those ideas is crucial: they must move other people to action or contemplation or both. Of course few in either the arts or the general public realize that this is how science works; few are aware that language is crucial, that brilliant visualizations, analogies and metaphor are vital and, even more important, that those devices are not there to be plucked off the lab shelf but must also be created by the scientists involved. Or at least this is the case when science is done well and expressed well. It was Crick, after all, who remarked that, "There is no form of prose more difficult to understand and more tedious to read than the average scientific paper."

Is there a course somewhere that, in the spirit of undergraduate English, focusses on a handful of brilliantly written and truly revolutionary scientific papers, examining both their scientific and artistic merit? If not, there should be. Call it "Watson, Crick and Shakespeare."

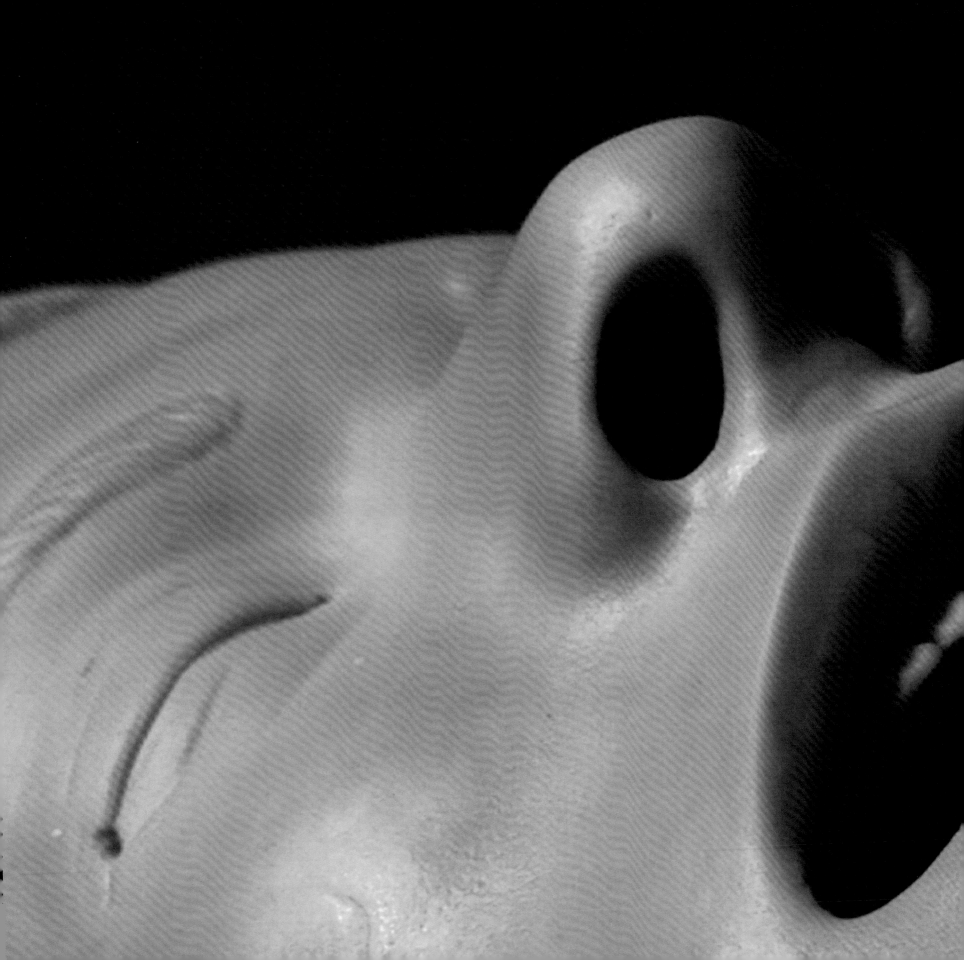

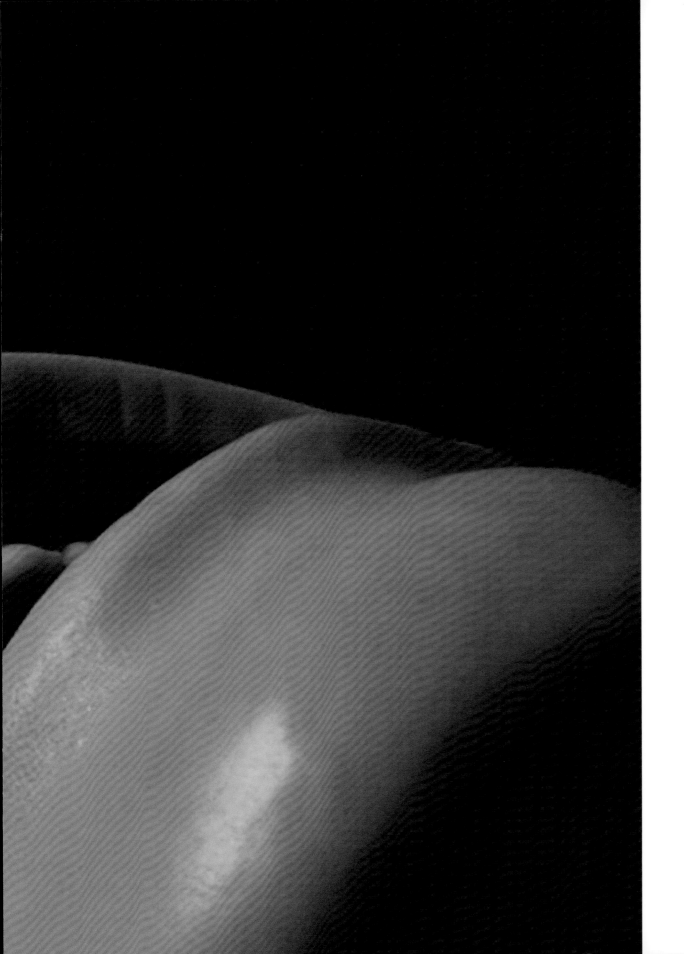

Christine Borland
SimMan (Face), 2007

Photograph
63 x 81.5 cm (unframed)
Courtesy of the artist and
Lisson Gallery, London
Photo: Dave Dunbar

SimMan is the registered name of a life-sized, computer-controlled mannequin designed for use by medical students as a training tool. Equipped with interactive technology that can generate automatic performance feedback, the surrogate human enables the simulation and treatment of various scenarios, allowing students to practise without fear of risk to patients. The use of such mechanized learning devices has undoubted benefits; however to increase our reliance on a virtual world, replacing the idiosyncratic and multifaceted reality of human life, raises grave questions.

BARTHA MARIA KNOPPERS

WHAT IS THE ROLE of policymaking in law where science intersects with informatics, where species mingle, and future generations can be altered? Is it just *Brave New World* enacted? Are there lessons to be learned from poetry for policymaking?

While seemingly incongruous, training in the art of poetry can be useful for the arduous and often inchoate world of policymaking. Taking as an example the period of surrealism that flourished in close parallel with World War II, its influence in both the artistic and political domains is still felt today. Providing as it did the tools for a counter-culture to established schools of thought, its influence and power surpassed its origins in Europe. It served to foster the independence of colonies from their colonizers. Embraced in South and Latin American, as well as in African art and literature (and later in Quebec during the Quiet Revolution), surrealism revolutionized and influenced both art and policymaking to adopt more authentic, dynamic and democratic approaches.

For example, in my view, available studies of Caribbean poetry demonstrate that when first subjugated, the colonized will show that they can write better, do better and surpass the style and the language of the colonizer. Yet, following this neo-romantic phase, it is surreal protest through art and poetry that forces the colonizer to listen to the colonized, and accept change. If however, the call for change stops at protest, the pattern of facile

acquiescence, and so of positivism, repeats itself until the next revolution. The patterns of policymaking in biotechnology follow a similar pattern to decolonization through artistic efforts. First, there is repugnance at the emergence of scientific breakthroughs, then protest, then acceptance by the authorities, then hopefully dynamic policy built on principles and vision—a creation that sincerely speaks to the individual and so to the universal.

Laborious and incoherent as this process may appear, its strength, like systems biology, is in its fluidity coupled with the desire to imprint values and vision on the framing of science for the next generation. In that sense, the more poetry or policy speaks to the individual who can recognize the "self," the more it is universal.

But law and poetry would only be tools—poor and paltry—if lost in political generalizations and ideologies. Passion is required for praxis. Where, then, comes the energy to truly transform? Not in party platforms, not in sound bites, but, in principles. There is poetry in the 1948 Universal Declaration of Human Rights. There is power and policy in *Hamlet*. Look at the lessons of the fall of the Berlin Wall and the tragedy of new walls being constructed—a testimony to our failure to use the passion necessary to achieve peace.

What, then, should we do today, or tomorrow, in this global world, in a global economy that lacks global solutions? Perhaps we need more not fewer principles, more not less faith, more not less energy to engage in this "surreal" epigenetic process. Is this idealist romanticism? Perhaps. But cynicism in the political system or regarding the transforming power of the arts is a poison that destroys, that feeds apathy—a concession to the lazy, the weak and the proud.

So we can learn from surreal poetry in developing policy, in moving from neo-romantic principles to practice, from protest to procedures. But then who is to say that we need not go from policy to poetry, from pragmatism to promise, from the personal to the universal? Perhaps therein lies the dialectic, the necessary dynamic of change.

Irrespective, we never know if we have truly taken the road less travelled. It does not matter. The only guidepost on any road is honesty and passion, to have a purpose and to be true:

> *Spaces between words, sense between lines*
> > *The language of each and own and them*
> > *Freedom unsquandered, a gift*

> *Endless platitudes on opportunity, on dreams*
> > *The road ahead, the choices offered,*
> > *Stillness in the minute, the day, a chance*

> *Will it be or when, unknown*
> > *Duty calls? No, created*
> > *The universal only personal when truly you*

Herein lies the intersection of policy and poetry, in the truly authentic and honest voice free of ideology and political manoeuvring. The truly personal voice can contribute to the construction of universal principles.

Superseding political, geographical and cultural divides, the very simplicity and humility of the single voice enriches both the complexity of international debate and the networks of public trust. Trust that the single voice, the single community or the single nation is respected and heard is necessary to the existence and meaning of international institutions. The individual, acting as world citizen, is a force to contend with. Policy built with the common good in mind resonates across boundaries and creates living, dynamic networks. Only then will policy be both personal and universal. Only then will it become poetry.

THE PLASTICIZED PREGNANT WOMAN AND
LEGAL RIGHTS OVER REPRODUCTIVE "MATERIAL"

TRUDO LEMMENS

WITH MY COLLEAGUE Lisa Austin, I co-teach a course called Privacy, Property and the Human Body, in which we discuss how the law, and particularly privacy and property rights, can help us deal with ethical issues associated with the human body. The course covers topics such as organ donation, DNA banking, the sale of gametes, treatment of human remains and decision-making over foetuses and embryos. We discuss the case-law, statutory rules and academic literature in law and bioethics. The debate can be vigorous, if most often academic. We contrast how different theories of law and different legal concepts treat these important questions. A free market approach to solving problems of organ or gamete shortage is, for example, put up against theories that emphasize concepts of human flour-ishing and human dignity. Notwithstanding the very concrete nature of the problems, they remain necessarily somewhat removed from real human experiences, in much the same way that life is generally molded in the court room into procedural and legal jargon.

In our many discussions, some of the hardest cases involve conflicts over the treatment of foetuses or embryos. They include situations where couples disagree about the disposal of frozen embryos; where a now brain-dead pregnant woman had previ-ously expressed a desire to be taken off life support, while her partner wants her body to carry a foetus to term; or where a pregnant woman's lifestyle risks damaging the foetus.

The relational component of the foetus (it still involves more than one person) and the claims about its special status unavoidably brings the touchy abortion debate into the classroom. The fact that the discussion remains generally focused on legal rights and privacy interests, and on interpre-tations of the Charter of Rights and Freedoms and Supreme Court jurisprudence, makes it less personal and seemingly more morally neutral. But discussing issues with obvious ethical ramifications in strictly legal terms leaves often a somewhat uncomfort-able taste; as if we're reducing ethical dilemmas to the black letters of a legal commentary without fully appreciating how these disputes unfold in the theatre of human life.

This is where art and imagery come in. Images can evoke a moral sensitivity and seem to connect us to deeper human experiences in a way that legal arguments and claims cannot. Even though images do not solve the disputes, and even though the emotive aspect of the imagery may not necessarily be universally shared, art can evoke the existence of an important "other" dimension that we tend to, and to some extent are pushed to, shy away from in legal debates.

Our course's session on human remains focusses on the question of whether individual self-determi-nation trumps familial and community interests in our bodies. We show slides of the *Body Worlds* exhibition to make the discussion more tangible.

This famous exhibition, which I would qualify as neither art nor science, involves the public showing of real but plasticized human corpses, set up in dramatic and sensationalist real-life postures. Many are opened up to expose the skeletal structure, as well as the workings of muscles, veins and internal organs when people are engaged in activities such as playing soccer or chess. Some students find even the more "innocent" images of real corpses creepy, disrespectful or outright offensive, but most think the decision over the use of one's body for such purposes is really a question of individual self-determination, that people should be free to decide what happens to their corpse.

Something very different happens when we put a picture of a plasticized pregnant woman on the screen. The woman is solemn, her belly opened up to reveal a five-month-old foetus in her womb. The silence that settles in the classroom at the sight of these two intermingled bodies is palpable. It's as if there is something indecent or intrusive about exposing this intimate relationship between an anonymous woman and her never-born foetus. Even though the woman must have agreed to have her body exposed and even though rights of privacy are only granted in law to people who are born alive, privacy and intimacy shared by the woman and her foetus seem to be breached here. The unease we feel at the sight of an unborn life within this womb reflects an intuitive and emotional recognition that there is something else going on here than the mere revealing of a woman's internal organs. Faced with

this image, most people seem to connect to another dimension of human existence that is not captured well by our legal discussions. We think of the woman who died before giving birth to this promise of life. We connect it to our own lived experiences. Verbal reasoning makes room for contemplative introspection. The emotional response we experience in that classroom is not unlike the connection we feel when a work of art touches us. We may not be able to express it precisely, but we feel a connection to deeper questions about our own existence and destiny and about our connectedness to others.

Showing these images doesn't necessarily change the classroom debate. In fact it silences it, at least temporarily. But silence speaks. The tangible emotions in the room drive home the message that this is not just about notions of legal right to self-determination and clashes of rights and interests. It is about real people having real dreams and hopes, being faced with challenges, expectations, joy and disaster. Even though thoughts may wander in different directions, we feel it's about human connectedness and interaction. It is about one independent woman, in a very intimate relation with a person-to-be and with unknown relations to others, who made a decision that not only affects her. Faced with this image, some in that classroom may add a spiritual or religious dimension to what they see and think about the sanctity of that relationship embedded in these strange plasticized "sculptures." Another student may reflect on his or her own experience with failed fertility treatment. Yet another student may feel discomfort or, on the contrary, relief by reflecting on a recent abortion experience. Another one may mourn in silence the recent loss of her foetus. Those with children ruminate perhaps about what they felt the first time they observed a foetus in their own or their partner's belly through an ultrasound. I, for my part, always think back about once holding a four-and-a-half-month old stillborn foetus in the palm of my hand and feeling a strange irrational guilt for abandoning this perfectly shaped tiny version of a boy in the hospital as "biological waste."

A deep connection and emotional understanding briefly enters the classroom. It binds us briefly, before we turn back to our academic quarrels, quickly forgetting, or at least hiding, the largely unconscious ways in which our interpretations of legal rights and obligations are always to some extent influenced by our own lived experiences and emotions, as well as our religious or moral beliefs that lurk in the background. Art and imagery allow us to reconnect to deeply felt emotional attachments and moral intuitions inside us. This dimension is inevitably subdued or censured in our discussions about how to organize societal relations and disputes through law. Uncensored artistic expressions can bring us back to the largely unconscious reasons why we feel strongly attached to some legal categories or approaches rather than others. Art reconnects our inner selves, perhaps all too briefly, with how we want to shape and organize our external relations through law.

Eduardo Kac

Lagoglyphs: The Bunny

Variations #8, 2007

Bichrome silkscreen print

15.7 x 21.2 in (40 x 54 cm) each

Edition of 50

Eduardo Kac

Lagoglyphs: The Bunny
Variations #10, 2007

Bichrome silkscreen print
15.7 x 21.2 in (40 x 54 cm) each
Edition of 50

Eduardo Kac

*Lagoglyphs: The Bunny
Variations #12, 2007*

Bichrome silkscreen print
15.7 x 21.2 in (40 x 54 cm) each
Edition of 50

Lagoglyphs is a series of twelve bichrome silkscreens created by Eduardo Kac in 2007 in which the artist develops a leporimorph or rabbitographic form of writing. As visual language that alludes to meaning but resists interpretation, the *Lagoglyphs* series stands as the counterpoint to the barrage of discourses generated through, with and around Kac's *GFP Bunny* (a genetically modified fluorescent rabbit that the artist created).

Eduardo Kac
Alba, the fluorescent bunny, 2000

LIANNE MCTAVISH

MOST VISITORS to La Specola, a natural history museum in Florence, are anxious to see its famous collection of anatomical waxes. Ignoring the displays of botanical and animal specimens, they head to the ten rooms that contain thousands of wax figures made between the end of the eighteenth and beginning of the nineteenth centuries. During that period, artists and anatomists dissected thousands of fresh corpses sent from the nearby Santa Maria Nuova hospital, observing them to create sculptures, first modelled in clay and then cast in wax mixed with resins and coloured pigments. The results continue to astonish spectators. Large glass cases feature life-sized human bodies that recline on embroidered cushions, while smaller cases present individual body parts. Labelled watercolour drawings line the walls around these exhibits, offering multiple views of the bones, muscles, heart, circulatory system, brain and reproductive organs.

Originally meant for teaching medical students about human anatomy, the wax models are no longer used for this purpose. Instead, they attract curious tourists, historians, and artists who draw from them. The primary appeal of the historical waxworks is their overt eroticism. The sculpted body parts are shown resting on cloth bags, indicating that they have recently been revealed to the viewer's gaze. As spectators move closer to peer at the exposed intestines and viscera, they are positioned as privileged voyeurs. This sense of voyeurism reaches its height in the reclining female figures popularly known as anatomical "Venuses." Here the bodies of nude female figures are opened to display glistening stomachs, livers and ovaries, with carefully rendered blue veins and golden fat deposits. The Venuses are adorned with long, sensuous hair and ornate jewelry. Their eyebrows, eyelashes and pubic hair are meticulously constructed from real human hair, and their delicate fingers curl to suggest that life once inhabited their bodies. Viewers are meant to see soft, warm and appealing flesh, not cold pedagogical tools.

Sex and science are united in the waxworks at La Specola in a way that may disturb modern viewers. According to historian Karen Harvey, eroticism was regularly associated with medical and scientific knowledge during the early modern period; even midwifery mannequins and obstetrical engravings could be used as bawdy forms of entertainment. By contrast, today's gynecology textbooks would excite few readers. They present the uterus in terms of mechanical function and focus on ovarian follicles rather than on whole bodies. Images of these follicles are often visually appealing, with bright dyes added for contrast, but, as Harvey points out, they remain unsexualized.

The link between desire and medicine nevertheless returns with a vengeance in contemporary popular culture. In the American television show *CSI: Crime Scene Investigation*, virtual simulations depict bodies that are penetrated, opened and defiled.

67

Another example is *Body Worlds,* the successful anatomical exhibition likely inspired by the waxes at La Specola. It includes human bodies preserved by plastination, a process invented by the German anatomist Gunther von Hagens. The recumbent figure of a woman who died while eight months pregnant is displayed in the format of a peep show: visitors must enter a curtained-off area before they can see how her skin has been pushed aside to expose an unborn fetus.

Many contemporary artists respond to this cultural phenomenon—promoting what Sigmund Freud would call the return of the repressed—by exploring the erotic potential of science and medicine. Sometimes they engage with the aesthetic appeal of scientific instruments, filling smooth glass jars with brightly coloured liquids. At other times they mimic medical methods of display, using pristine glass cases to produce the value of the objects enclosed. They stage sterile laboratory environments to hint that something sinister is occurring just beneath the surface. They fabricate and then give undue attention to prosthetic limbs. They use graph paper to produce diagrams that feature forms of organic growth rather than measured precision. In all of these examples, artists grasp the contradictory nature of modern medicine, which is driven by both reason and passion. They also reveal the obsessive aspects of much contemporary science.

In the end, these artists invite viewers to recognize their own erotic investment in modern science and medicine, while encouraging scientists to embrace the creative and embodied elements of their own work.

ERIC E. MESLIN

The great instrument of moral good is the imagination.

—PERCY BYSSHE SHELLEY (1821)

HOW DOES SOMETHING capture one's imagination? I first became interested in bioethics thirty years ago when an enlightened high-school biology teacher in Toronto asked his students to select a biological process and technology and study them for a year. I selected infertility and the as-yet-to-be-perfected technology of *in vitro* fertilization. Shortly after I handed in my final assignment I was stunned to learn of Louise Brown's birth, the world's first test-tube baby, gradually learning that the most exciting part of the story was not the scientific technique itself but the ethical issues it raised for society. And so began my own personal travels of imagination regarding and wonder at the impact of science and technology on society.

Bioethics is the study of the ethical, legal and social issues that arise in health care and science. Its practitioners—philosophers, lawyers, theologians and others—have prided themselves in being able to imagine the future and offer thoughtful insight into the moral challenges that society could face as technology speeds along the highway of social progress. In this way, bioethics has been a special type of social commentator: in part looking into the future of science and offering insight into the prospect for human health and human morality and, at other times, trying to make sense of recent developments. Bioethical imagination involves both attributes, and an entire cadre of specialists have sprung up to formally engage in the "academicization" of wonder about science and technology. Their contributions have been significant—by helping the public and policy-makers to understand the ethical implications of science.

But academics were not always granted this privileged status with respect to morality and ethics. The early eighteenth-century Estates of the Realm (clergy, nobility, bourgeoisie) reflected the dominant role of the clergy (and to some extent the learned nobility) as authoritative arbiters of public morality. The tradition of establishing an intellectual elite to provide guidance and advice continues today. In 1997 U.S. President Bill Clinton responded to the dramatic news of the cloning of Dolly by writing an executive order prohibiting any federal funds to be used for human cloning. He asked his National Bioethics Advisory Commission to conduct a review of the ethical and scientific issues and "report back to me in 90 days." Political actions of this kind are especially strong expressions of moral sentiment that may be as much intended to represent the unspoken view of the people as to stake out political territory. Ironically, legislation has yet to be passed in the United States banning human reproductive cloning. The American public overwhelmingly opposes human reproductive cloning, but politicians (on both sides of the aisle) worry that passing legislation might have unintended political consequences. Canada and other countries have joined these discussions by attempting to legislate morality on matters of science and have found it difficult. A majority of the public supports embryonic stem cell research, and yet limits are placed on the extent to which the government will fund such research.

It strikes me that political or academic expressions of social response are necessary but not sufficient means for fully understanding and appreciating the ethical implications of science and technology. Important as the contribution of bioethics may be to turn its analytic and normative gaze on moral problems in science and technology, it does not have an exclusive claim on this activity. We have always known that the arts community can engage the public to consider the ethics of science and technology whether through movies, books, or plays and other media. Perhaps, just perhaps, Percy Shelley was on to something.

Liz Ingram & Bernd Hildebrandt
Perplexed Realities 1 (computer
rendering of proposed on-site
installation at the Art Gallery of
Alberta), 2008

410 x 366 x 10 cm
Dye sublimation digital output on fabric,
Plexiglas rods, vinyl type
Courtesy of the artists
Photo: Bernd Hildebrandt

Perplexed Realities 1 and 2 is an installation that utilizes text and image combinations in order to question the impact of emerging biotechnologies upon natural environments and our bodies. In particular, the artists are interested in exploring the long-term ramifications of technology on our relationship to forces and cycles in ecosystems, specifically in relation to water preservation, pollution and biodiversity. To pursue these creative questions Liz Ingram and Bernd Hildebrandt have digitally printed layered images of water, human chromosomes, poetry and the body onto large fabric screens. These are placed in an open gallery environment in a manner that confronts the viewer.

Liz Ingram & Bernd Hildebrandt
Perplexed Realities 2
(working image), 2008

410 x 366 x 10 cm
Dye sublimation digital output on fabric,
Plexiglas rods, vinyl type
Courtesy of the artists
Photo: Bernd Hildebrandt

PETER W.B. PHILLIPS

POLITICS makes strange bedfellows. At root, politics is about the pursuit and use of power, either to achieve some positive objectives or to block and limit the objectives of others. Art has played a critical role in visualizing and popularizing our conceptions of our realities and our opportunities. In this way, art often reflects the goals and issues of the day and frames our views and responses to the existing distribution of power.

In the context of new technologies, there are always proponents of change—governments seeking more power and glory, industry and entrepreneurs seeking greater wealth, and social groups seeking more diverse and personal goals—who are always confronted with alternate views and goals that are often expressed by artists and their art.

Art, in all its forms, has played an integral role in politics since the beginning. Musicians, painters, sculptors, cartoonists, writers, playwrights and filmmakers have all either explicitly or implicitly contributed to the public discourse, by flattering, observing or critiquing the leaders and their approaches to the issues of the day. Meanwhile, economic, political and social leaders have always called on, conscripted or commissioned artists to portray them or their goals and accomplishments in a positive light. Most of our earliest art, from Mesopotamia, Central and South America, and Egypt depicts rulers as powerful and beneficent, pursuing grand goals, such as conquest, discovery or just good governance.

As time has passed and civilizations have evolved, the artist's role in society has expanded, as they have grown beyond the patronage of elites. While often impoverished and unappreciated in their own times, artists have come to occupy a privileged and enduring place in society, helping us to define our conceptions of truth, beauty and justice, and mobilizing ideas and people to support or usurp powerful actors and systems.

Renaissance art defined many of our concepts of human beauty—da Vinci's *Mona Lisa*, Botticelli's *The Birth of Venus* and Michelangelo's *David* rediscovered and promoted the notion of the human as beautiful. Their conceptions of beauty and proportion resonate today in the art and culture of many countries in the twenty-first century. Then the Romantic movement defined our conceptions of nature—Constable's pastoral paintings still encompass for many the ideal and pure "nature" that modern industry and man despoil. More recently, modern and postmodern art is credited with symbolizing and crystallizing people's sense of alienation and disconnect from many of the existing economic, social and political power systems that prevail today. Experimentation with new styles (cubist, abstract), new media (industrial materials, cathode rays, body excrement) and new concepts (installation or performance art, involving the viewer as part of the art) has touched different sensitivities and highlighted new aspects of our traditional systems.

Against this backdrop of the artist as a mirror for our dreams, follies and foibles, art and artists are now beginning to explore the deeper meaning of the recent discoveries and experiments with our genomes, and the genomes of other species and organisms that live among us. While artists have been able to capture the elegance and beauty of the double helix DNA strand, the corresponding knowledge and our capacity to use it presents a major challenge for artistic treatment.

In one sense, this technology enters the realm of the ethereal, being too small to see and too diverse in its expression to easily conceptualize. Thus, it is harder for artists to represent these recent advances in knowledge and science than they could with the mechanical innovations that precipitated and characterized the Industrial Revolution. Art in that earlier period of economic and social revolution was able to define the awe and might of the new innovations and to envisage and record both the great wealth and widespread poverty generated by its application.

The political debate about new technologies has already been influenced by the artistic conception of this new knowledge. Artists have taken the concept of genetic engineering and begun to characterize the futures it offers. On the one hand, some artists envisage a golden age of development, where truth, beauty and justice prevail, and wealth, health and abundance banish poverty, disease and want. The photo collages on the covers of *Nature* and *Science* in February 2001, announcing the sequencing of the human genome, solidified in people's minds the

link between the invisible genome and the beauty of human diversity. The natural symmetry of the double helix has almost made it an icon in modern art and culture. Alternatively, some see the reverse image of this world through a metaphoric Dorian Gray mirror, with uniformity and conformity driving out diversity and dissent. Adam Brandejs' *Genpets*, shown in an exhibit at the Montserrat College of Art in Beverly, Massachusetts, and the Experimental Art Foundation 2007 Exhibitions Program in Adelaide, Australia, offer glimpses of the potentially ugly underbelly of the technology. In this way, artists have begun to capture the essence of the political debate that surrounds and underpins this latest technological revolution.

The synthesis of the emerging artistic interpretation of the genomics revolution and the related economic, political and social expectations and goals for the technology are likely to produce novel hybrids that will influence society and our environment for generations to come. This technology has the potential to move us further from the Hegelian historical process, replacing human progress with ad hoc, materialistic innovation. As our ability to use our new genomic information advances, we are presented with opportunities to build within living organisms both new forms and functions. We have already entered an era of designing organisms for lifestyle purposes (blue carnations and roses and most recently pets, such as the Ashera cat, a cross between the African Serval, the Asian Leopard Cat, and a domestic breed).

More fundamentally, however, biotechnologies offer the opportunity to build in biological or other novel control systems to supplement or replace our human governing systems. In many of these cases we are combining these new forms with visually or conceptually appealing attributes—genetically modified crops and bacteria have been designed to fluoresce in attractive greens or blues when facing abiotic stresses or in the presence of toxic elements, such as unexploded land mines or industrial efflu- ents. The deliberate integration of form and function has the potential to further dehumanize our governing systems, and to alienate humans from the natural world.

JAI SHAH

AS A FIRST-GENERATION CANADIAN whose parents and grandparents were born in East Africa and India respectively, I remember being fascinated as a child by art and sculpture brought back from various trips visiting relatives—the different representations of the human form, and the features that changed or remained similar across these geographies. But these were far from my mind as an undergraduate studying biological sciences in the mid-1990s, when genetics was in ascendance and the Human Genome Project gathered steam and controversy.

At the time, genetics was known by us students for its "eureka" moments, of which there were two generally accepted sorts. The first were the culturally recognizable, historical moments—the discovery of the double helical structure of DNA or the moment when the first draft of the human genome sequence was published. The second was a unique moment of understanding that we all experienced as an epiphany of sorts—the instant when one made that leap from simply studying DNA and RNA to using genetics as a powerful analytical lens onto a host of biomedical issues. Genetics was no longer about the structure of molecules: it was about information flow and its regulation, and this shift was significant enough to be a topic of regular conversation.

What we didn't recognize at the time was that genetics could be even more than this: not just a scientific body of knowledge, a tool, or even an analytical lens but a subject of analysis in its own right. The realization that genetics, too, could be critiqued represented, for this author at least, a third such moment. Genetics was suddenly linked to questions of discrimination, race and ethnicity, and human nature itself. Watson and Crick's double helix model, the bar codes of gel electrophoresis, the anthropomorphized innocence of Dolly the cloned lamb's gaze, to name a few images that burst to mind, all resonated not just as celebrated feats of scientific discovery but as modern cultural icons. As a result, this student of genetics now finds it nearly impossible to describe the scientific significance of biotechnology without explicit reference to its socio-cultural significance. Fact and interpretation became inextricably linked.

. . .

This linkage became impossible to ignore when I reflected on how pervasive the images related to genetics have become. This is a particularly contentious issue, for there is a concern that using imagery and metaphor to communicate science and scientific findings is, although perhaps inevitable, also fraught with the danger of people coming to inaccurate assumptions, understandings and conclusions. Nowhere is this more true than in genetics, whose practitioners know that its erroneous interpretation has led to consequences ranging from employment discrimination to ethnic cleansing.

So every time a metaphor is used by scientists on the front lines of discovery, other scientists and social commentators hasten to point out that metaphors may "miss the real story." Some of these powerful metaphors include: genes are "blueprints" and "recipes"; the institutionalization of genome science through the Human Genome Project is "the holy grail"; DNA is a "master molecule" of information, with the cell a secondary "machine" carrying out its instructions.

Each metaphor serves to translate the miniscule, esoteric and remote world of molecules into something concrete, familiar and accessible. But genes are *not* blueprints. And DNA is *not* a self-replicating "master molecule." DNA requires other proteins and systems to make copies of itself, and cellular function is determined by an interaction of genes, environmental factors and random events. As Richard Lewontin noted in *Science*, "the use of metaphor carries with it the consequence that we construct our view of the world, and formulate our methods for its analysis...as if the metaphor were the thing itself." The use of metaphor in science, in other words, is to be handled gently and delicately.

Yet if a metaphor can lead to errors of understanding about the workings of the molecular, cellular and biological, the same metaphor may prove useful for other vehicles. Marc Quinn's 2001 portrait of Sir John Sulston, the Nobel laureate who led the UK contribution to the human genome sequence, is billed by London's National Portrait Gallery as "a detail of Sulston's genome—the 'recipe'

to make him." A framed set of opaque beads, each a colony of bacteria with an inserted fragment of Sulston's DNA, the portrait provocatively claims to "precisely capture what is unique about" its subject. Quinn uses the "recipe" metaphor to challenge our notions of what makes us human, the ways in which we are similar or unique as individuals, and the personal impact and implications of genome science. Unlike social commentators, imagery and metaphor are important, even essential and perhaps inherent, to the artist's critique.

■ ■ ■

And so we have come back to art, and to exhibits and compilations like the book containing this essay. For this student, understanding genetics as a social institution with norms and values was a conceptual leap, one that immediately translated into material changes in his interests and research. Does the conceptual similarly spark the subject matter for artists, or vice versa? Do artists feel that the increasingly visual nature of genetic information (and perhaps science in general), and its iconic stature, has contributed to its prevalence in artwork?

If scientists, social commentators and artists share similar goals—to critique knowledge and practice—then is it not peculiar that their approach to that most inevitable concept, the metaphor, is so dramatically different? If critics recognize the value of imagery, analogy and metaphor in art, then how do they distinguish these metaphors from the ones they caution scientists against? Art uses imagery to find new ways of seeing, understanding and critiquing the world, and metaphor allows art to stand out as an example of how genetics is subject and not just tool.

Now just as art has taken on questions of biology, technology and human nature through metaphor, it has also been embraced and institutionalized as a vehicle for communication by scientists, lay and specialist publications, and even funding bodies. Vendors offer to create a unique "DNA portrait" based on gel electrophoresis, involving purchasers in the creation of the work through a simple mail-in cheek swab. Large exhibitions tour the country and are uploaded onto the Internet. They are often accompanied by essays hoping to capture why *this* particular moment in time is uniquely suited for, and worthy of, critical engagement.

...Is it?

Catherine Richards

Charged Hearts, 1997

Site-specific work

Brass, copper, glass, electronics,

electrodes, electron guns, computer,

luminescent gases, lights

42 x 83 x 60 in

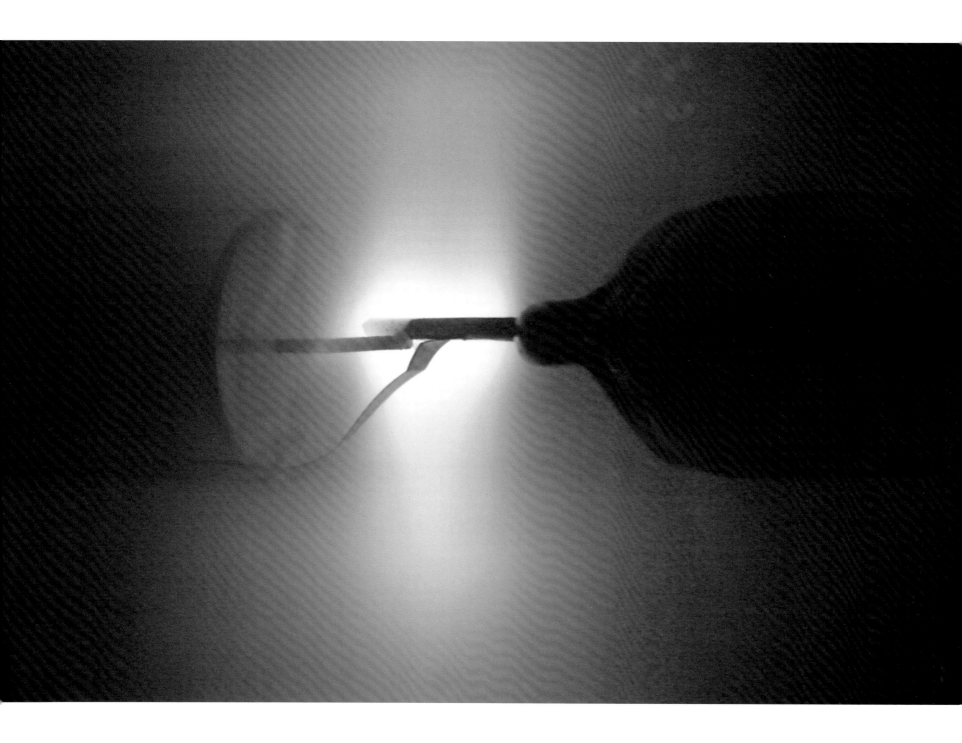

Photos: Daniel Gamache/NRC Photo
(p. 81), Mitch Lenet (pp. 82–83), and the
National Gallery of Canada (p. 84)

Our technological culture has created an environment of electromagnetic signals, pagers, television, fluorescent lights, and voltage lines, to name a few. As our technology becomes wireless the space around us is filling up. It is not limited by our body boundaries. We are already plugged in, all the time, systemically bonded.

The human heart, the symbolic seat of emotions, is also one of the body's better-known electromagnetic fields. The electromagnetic wave is the heartbeat itself. Hearts are charged both literally and figuratively. We can say "the hearts excite" and mean that the electrons are firing as much as we mean that our hearts are moved.

Charged Hearts is a site-specific work. It plays with the prescribed behaviour that is the norm of museum and gallery spaces: "Do Not Touch." Spectators transgress, by stepping on the glass floor, by reaching into the cabinet and by picking up the object. By stepping up on the glass the spectators are on display. By entering the cabinet they have crossed the boundary separating themselves from the artist's work. By picking up the object they plug in.

The spectator "plugs in" by lifting a jar, thus exciting a glass heart. Its phosphorescent gases then beat luminescently. Situated in the middle cabinet, the "terrella" (small world), simultaneously excites, creating an aurora borealis, like a cathode ray tube stripped bare.

Charged Hearts was commissioned by the National Gallery of Canada, 1997.

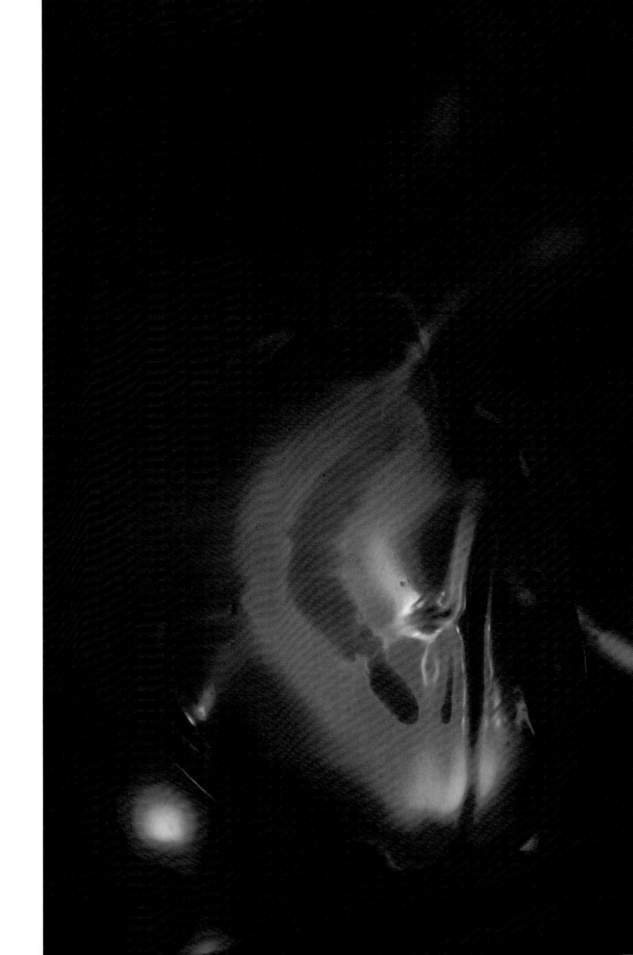

Charged Hearts at the National
Gallery of Canada, 1997.

AIN'T DAT-A BEAUTY

STEPHEN STRAUSS

I WRITE ABOUT the relationship between art and science profoundly, nay even fatally, disadvantaged by understanding both much and little about each of them.

I am neither an artist nor a scientist, but my wife is a painter and over the decades I have watched her paint and questioned her about how she did it. As far as I can tell, her method might be called the Antichrist of the scientific method. She is not even a nano interested in the reproducibility of her results, in the validation of her suppositions by data, or in an objective unveiling of the natural world numerically.

She feels and then tries to express her feeling with colour and line. Period. Or more appropriately, given the passion of her work: exclamation point. I have never encountered anyone who has responded to one of her paintings as if it were a formula or a statistical expression of reality. Rather, they observe her visual depiction of feeling and they feel back. On occasion, they buy her paintings and put them up in their houses and every day observe her painted feelings and merge them with their unpainted ones.

Conversely I recently spent a busy hour in a large particle physics laboratory where you need millions of dollars and hundreds of people to discover scientific truths that only manifest themselves as jumping dots and bouncing lines on a computer screen. Nobody in the lab made any reference to "beauty" in the construction of an experiment. Nobody mentioned how "feelings" shaped their work.

In retrospect, the lab fit my notion of the place where boredom goes to be bored. Its methodological mantras were: repeat; repeat; re-analyze data; run the tests again; discard what you think, but don't know for sure, are false positives; repeat; repeat. Then, when you think you have found something true about the universe, you present this highly quantified, typically exceedingly narrow, verity to others and request they point out where your repeats derailed. If they find nothing amiss, well, mini-hurrah! And if they do find something wrong, it's back to searching for another way to use boredom's own boredom to unveil nature.

Furthermore, the form scientific communication takes is—and I leave you parse almost any scientific paper to verify this—data truth without data beauty. Indeed, scientific aesthetics remain subject to Huxley's breezy remark that "science is organized common sense where many a beautiful theory was killed by an ugly fact." What might be attached to this could be that while a given fact may lack context or connection with any consistent theory, that's more akin to loneliness than ugliness.

Artists, as far as I can tell, almost never willingly submit their work to other artists for approval. In fact, they often disdain the opinions of others, since what they do is supposed to be *sui generis*, who but the artist can judge his or her art's worth? I mean, what do buyers even know, not to mention rancid, jealous, idiotic art critics?

Given all this, what I find quite amazing is the notion that artists and scientists really would have much to say to one another in any realm. While there sometimes seems to be a conversion in the realm of originality, even that seems to me a miasma.

The best artists and the best scientists are each vexingly creative. They discover new things either in themselves or in the world, and their creativity is always wondrous to those who wish to be creative themselves. But surely creativity is a sensibility that bounds about in every human activity; businessmen can be creative, as can athletes, generals, carpenters and, no doubt, certain sex workers. If creativity is how art and science are similar, then these two areas of human activity are also linked, inspiration by inspiration, to almost every other kind of human action.

Ultimately we must accept that art and science are not just different but that there is almost a fundamental opposition inherent to their approaches. Each is the other's methodological Antichrist, and, in that, the most amazing thing about the relationship between art and science is that the human brain can absorb and believe in replicable, numerical, objective expressions of reality, as well as figurative, non-numerical, subjective expressions of the same world, all without losing consciousness or going briefly mad. It is a *Homo sapiens* aptitude that demands to be recognized in writing, not by using a period or an exclamation point but by something more—a wow followed by a WOW.

ART AND SCIENCE
Genomics and Food

MICHELE VEEMAN

FOOD is a topic of almost universal interest, as well as being the subject of inspiration and a focus of professional endeavours for both artists and scientists. The compelling cultural context and the importance of the social implications of food have been depicted by many artists. One example, viewed in the mid-1700s through the eyes of Peruvian artist Marcos Chapata, is seen in his painting of the Last Supper, which hangs in the Cusco Cathedral. This shows Christ and his apostles seated at a table on which local Peruvian foods, including peppers and papayas, are arrayed; a central platter holds a cooked guinea pig (a staple food in the Peruvian highlands). Locals suggest that the drink depicted to accompany this meal is the maize-based beer, *chicha*. Canadian painter Mary Pratt provides a rather different view of the social connotations of food in her celebrations of its domestic focus. Her food paintings include images of a hanging moose carcass and fish or chicken ready to be cooked, as well as luminous depictions and reflections of homemade jelly and various fruits.

In contrast to the work of those artists who depict food, scientific focus on food includes the work of food scientists, who have sought for ways to preserve and process food and to improve food safety in the context of the increased distances and longer periods of time that separate food producers from consumers in today's modern society. The modern food scientist's focus becomes increasingly important to consumers as most of the world's population now lives in urban areas and few of us have the time, desire or ability to prepare foodstuffs from the very basic forms in which they are harvested. As well, plant and animal geneticists, and plant breeding specialists and agronomists, amongst others, have sought plant and animal varieties that are both more productive and better able to withstand disease. This reduces the likelihood that continually evolving plant and animal diseases will lead to food emergencies, such as occurred in the catastrophic years of widespread Irish potato blight (1845-1848) in which crop disasters, accentuated by political and economic mismanagement, led to famine, death, eviction and the emigration of a large percentage of Ireland's population.

The methods used by food scientists have evolved since Mendel's seminal work, based on his studies of plants (specifically the trait inheritance of peas), later credited as forming the basis of modern genetics. Crop selection in Canada has emphasized improved frost tolerance, yield and grain quality. Much current work by plant scientists is directed to improvements in crop plants that will withstand environmental pressures associated with climate warming, such as increased heat or the stress of less abundant rainfall and the ability to grow in degraded or saline soils. (Increased emphasis in plant breeding is also being directed toward potential biofuel production.)

More recently, plant and animal scientists have been able to call on the tools of molecular biology to identify and select for genetically determined traits associated with disease and pest resistance, faster growth and improvements in the nutritional characteristics of particular foods and animal feed. Advances in the fields of genomics have considerably extended the tools available to agricultural science: new, cheaper methods of screening genetic material have made it possible to search more rapidly for the determinants of desired or undesired traits (for example to discover whether or not and where these components are available within the population of a particular type of plant or animal) and to identify molecular markers of such traits, in order to facilitate marker-assisted selection. These tools are making it possible to considerably extend and speed up selection while still using conventional plant and animal breeding techniques. However, agricultural biotechnology, involving the use of transgenic methods—sometimes referred to as genetically modified (GM) or genetically engineered (GE) crops— viewed as a non-conventional or novel technology, has been the subject of controversy and debate.

Nonetheless, farmers are quick to adopt new seeds, breeds and methods where these improve their livelihood. Consequently, outside of Europe, during the last decade there has been widespread adoption of some GM/GE crops: as with corn and soybeans in the United States, soybeans in Brazil and Argentina, canola in Canada, and cotton in the United States, Australia, China, India and many other nations. Most recently, increases in GM/GE crop production in the developing world have led these types of crops to account for a growing proportion of the world's cropped area.

The need for bio-safety assessment and necessary regulation is uncontested. But many agricultural scientists are concerned that regulatory requirements and capacity, together with the uncertainty associated with polarized views of agricultural biotechnology, may reduce the ability for use of agricultural biotechnology to contribute to improvements in the quantities and qualities of basic staple crops that are important to feeding the hungry and improving the incomes of the poor in developing nations.

The cultural and social context of our lives typically determines whether a particular food is desired, accepted or rejected, whether it is the guinea pig meat so prized in the Andes, the moose meat some Canadians relish, or the GM/GE food that some shun. Science increasingly affects the types of food available in modern society, and although scientists must generally work within their social context, artists often challenge our views of culture and society. Art may also interpret who we are at any given moment in our collective history, including the food we put into our bodies. Our tastes may change as we are exposed to different cultures, to changes in social influences and to changes in products that have been impacted by science, but there is no doubt that we can view the food we eat as a kind of mirror reflecting where we came from and the times in which we now live, a mirror often held up by the artist.

ATOMS AND BASE PAIRS
The New LEGO

GREGOR WOLBRING

WHEN I WAS YOUNG I enjoyed playing with LEGO, the ingenious plastic bricks I could link together to generate and create objects, art—there are so-called LEGO artists—in unending variations. Synthetic biology is a recently emerged scientific field, and molecular manufacturing a field expected to mature soon; both make me feel as if I am playing LEGO all over again.

The field of synthetic biology could be defined as a bottom-up approach to designing life from the base-pair level. Synthetic biology, according to one definition on the synthetic biology research community webpage is: (a) the design and construction of new biological parts, devices and systems; and (b) the re-design of existing, natural biological systems for useful purposes. It uses, among other things, BioBricks, a standard for interchangeable parts, developed for building biological systems in living cells.

In the end, one builds, designs, creates life from the base-pair level up. Craig Venter, the DNA researcher who was involved in the race to map the human genome (some say he won the race), and his team have built a synthetic chromosome from the base-pair level. In this case, according to the *London Guardian*, the synthetic chromosome consists of 381 genes that were generated by putting together 580,000 base pairs of genetic code. These base pairs are the LEGO bricks. Put together, they can generate all kind of genes. So far, the field of synthetic biology uses mostly the traditional four bases, which are the

foundation of the human genome. However, artificial bases that can pair with the traditional four are already available, allowing for genes, genomes and chromosomes with characteristics not existing in living matter today.

Molecular manufacturing is in some ways the non-living matter equivalent of synthetic biology. It's another bottom-up approach. The anticipated field of molecular manufacturing puts together atoms to design, create and build non-living matter. Many might be aware of the food replicator in all Star Trek films where one says, for example, "Coffee," and the machine builds a cup of coffee, synthesizing the beverage molecule by molecule. The term "nanotechnology" was actually used first to describe a way to manufacture something from atomic molecules. Only in recent times, with the shift to nanoscale science, has the meaning of nanotechnology evolved to where it is now.

Now what has the above to do with arts? For one, the bottom-up design of life and non-living matter could be seen as art is part of using LEGO bricks. Furthermore one can anticipate—taking into account the long history science and art have together—that the arts will be employed to engage the public in understanding synthetic biology and molecular manufacturing, in much the same way that arts have been used to engage the public on other scientific issues. Artists are embedded in scientific endeavours, and artists increasingly use scientific material for their own purposes (e.g., the Tissue Culture & Art Project) and as a means to generate a work of art. It seems logical to expect that artist will use base pairs and atoms as material when they become available to them.

But the question is this: Does the role of the artist stop there? I participated in various science and arts events at the Banff Centre involving nanotechnology, biotechnology and health research. These events not only highlighted how science can be used by artists, and how artists can be used by the sciences, but they also investigated the role artists play or should play in the governance of science in general. Do they have an obligation to the public to think about the usage of science as artistic material? With regards to using atoms and base pairs, it is reasonable to expect that artists will face a discourse regarding whether or not there should be a limit to artistic expression, in much the same way that scientists will face questions about the limits of synthetic biology and molecular manufacturing.

Do artists have to take into account the social consequences of using science as artistic material? Should artists play a role in the governance of science and technology? If yes, what role? If no, why not? The more artists use science and science uses artists, the more vital it is that artists try to answer these questions. Using base pairs and atoms might feel like playing LEGO all over again, but the consequences and implications are rather more serious.

Through the use of digital images, performance, installation, as well as biotechnology, artists Shawn Bailey and Jennifer Willet create innovative images and environments that challenge viewers to consider numerous ethical and conceptual issues related to emerging biotechnologies. The creative strategies driving this work are multifaceted and include the creation of a fictional biotech corporation, which, ironically, critiques the impact of commercialization on public perception of the biomedical industry. Bailey and Willet's work also questions the location and nature of discourse surrounding emerging biomedical technologies by transforming gallery spaces into functional laboratory environments, which shifts a viewer's relationship to this technology and which encourages broader public dialogue. In the creation of BIOTEKNICA, Bailey and Willet spent extensive time working in biology research laboratories around the world, exploring the artistic applications of cell lines, tissue culture and tissue engineering technologies.

SUGGESTED READINGS

Andrews, Lori. "Tissue Culture: The Line Between Art and Science Blurs When Two Artists Hang Cells in Galleries." *The Journal of Life Sciences* (September 2007): 68-73.

Avise, John C. "Evolving Genomic Metaphors: A New Look at the Language of DNA." *Science* 294, no. 5540 (2001): 86-87.

Beyleveld, Deryck, and Roger Brownsword. *Human Dignity in Bioethics and Biolaw.* New York: Oxford University Press, 2001.

Bonnicksen, Andrea L. *Crafting a Cloning Policy: From Dolly to Stem Cells.* Washington, D.C.: Georgetown University Press, 2002.

Caulfield, Timothy. "Popular Media, Biotechnology and the 'Cycle of Hype.'" *Journal of Health Law and Policy* 5 (2005): 213-33.

Caulfield, Timothy, and Roger Brownsword. "Human Dignity: What Does It Mean in the Biotechnology Era?" *Nature Reviews Genetics* 7 (2005): 72-76.

Cohen, Cynthia B. "Creating Human Nonhuman Chimeras in Stem Cell Research." In *Renewing the Stuff of Life: Stem Cells, Ethics and Public Policy.* New York: Oxford University Press, 2007.

Cohen, Cynthia B., et al. "Oversight of Stem Cell Research in Canada: Protecting the Rights, Health, and Safety of Embryo Donors." *Health Law Review* 16, no. 32 (February 2008): 86-102.

Conner, Clifford D. *A People's History of Science: Miners, Midwives, and "Low Mechanics."* New York: Nation Books, 2005.

Dawkins, Richard. *River Out of Eden: A Darwinian View of Life.* New York: Basic Books, 1995.

Doolittle, W. Ford, and Eric Bapteste. "Pattern Pluralism and the Tree of Life Hypothesis." *Proceedings of the National Academy of Sciences of the United States of America (PNAS)* 104, no. 7 (2007): 2043-49.

Dworkin, Ronald. *Life's Dominion: An Argument About Abortion, Euthanasia, and Individual Freedom.* New York: Knopf, 1993.

Edelman, Murray. *From Art to Politics: How Artistic Creations Shape Political Conceptions.* Chicago: University of Chicago Press, 1996.

Edwards, David. *Artscience: Creativity in the Post-Google Generation.* Cambridge, MA: Harvard University Press, 2008.

Einsiedel, E., ed., *Emerging Technologies: From Hindsight to Foresight,* UBC Press (forthcoming 2008).

Greely, Henry T., et al. "Thinking About the Human Neuron Mouse." *American Journal of Bioethics* 7, no. 5 (2007): 27-40.

Hardcastle, Rohan J. *Law and the Human Body: Property Rights, Ownership and Control.* Portland, OR: Hart Publishing, 2007.

Harris, John. *Enhancing Evolution: The Ethical Case for Making Better People.* Princeton, NJ: Princeton University Press, 2007.

Hart, Jonathan, ed. *City of the End of Things: Great Minds on Civilization and Empire.* Toronto: Oxford University Press, 2008.

———. *Interpreting Cultures: Literature, Religion and the Human Science.* Basingstoke, New York: Palgrave Macmillan, 2006.

Harvey, Karen. *Reading Sex in the Eighteenth Century: Bodies and Gender in English Erotic Culture.* Cambridge: Cambridge University Press, 2004.

Heiferman, Marvin, and Carole Kismaric. *Paradise Now: Picturing the Genetic Revolution.* New York: Distributed

Art Publishers (Tang Museum and Skidmore College), 2001.

Hui, Jingyi, Maryline Mancini, Guangdi Li, Yuan Wang, Pierre Tiollais and Marie-Louise Michel. "Immunization with a plasmid encoding a modified hepatitis B surface antigen carrying the receptor binding site for hepatocytes." *Vaccine*, 17, nos. 13–14 (January 1999): 1711–18.

Kac, Eduardo, ed. *Signs of Life: Bio Art and Beyond*. Cambridge, MA: MIT Press, 2007.

———. *Telepresence and Bio Art: Networking Humans, Rabbits and Robots*. Ann Arbour, MI: University of Michigan Press, 2005.

Karpowicz, Phillip, Cynthia Cohen, and Derek van der Kooy. "Developing Human-Nonhuman Chimeras in Human Stem Cell Research: Ethical Issues and Boundaries." *Kennedy Institute of Ethics Journal* 1, no. 2 (2005): 107–34.

Kay, Lily. *Who Wrote the Book of Life? A History of the Genetic Code*. Stanford: Stanford University Press, 2000.

Korobkin, Russell. *Stem Cell Century: Law and Policy for a Breakthrough Technology*. New Haven: Yale University Press, 2007.

Lakoff, George, and Mark Johnson. *Metaphors We Live By*. Chicago: University of Chicago Press, 2003.

Leathers, Howard D., and Phillips Foster. *The World Food Problem: Tackling the Causes of Undernutrition in the Third World*. 3rd ed. Boulder, CO: Lynne Rienner Publishers, 2004.

Leavis, Frank Raymond. *Two Cultures? The Significance of C.P. Snow*. New York: Pantheon Books, 1963.

Lemmons, Trudo, and Duff R. Waring, eds. *Law and Ethics of Biomedical Research: Regulation, Conflict of Interest and Liability*. Toronto: University of Toronto Press, 2006.

———, Roxanne Mykitiuk, and Mireille Lacroix. *Reading the Future? Legal and Ethical Challenges of Predictive Genetic Testing*. Montréal: Éditions Thémis, 2007.

Lewontin, Richard C. *Biology as Ideology: The Doctrine of DNA*. Toronto: Anansi Press, 1991.

Massey, Lyle. "Pregnancy and Pathology: Picturing Childbirth in Eighteenth-Century Obstetric Atlases." *The Art Bulletin* 87 (2005): 73.

Nelkin, Dorothy, and M. Susan Lindee. *The DNA Mystique: The Gene as a Cultural Icon*. New York: Freeman, 1995.

Peat, F. David. *From Certainty to Uncertainty: The Story of Science and Ideas in the Twentieth Century*. Washington, D.C.: Joseph Henry Press, 2002.

Phillips, Peter W.B. *Governing Transformative Technological Innovation: Who's in Charge?* Oxford: Edward Elgar Publishing, 2007.

Rappoport, Leon. *How We Eat: Appetite, Culture and the Psychology of Food*. Toronto: ECW Press, 2003.

Scott, Christopher Thomas. *Stem Cell Now: From the Experiment That Shook the World to the New Politics of Life*. New York: Pi Press, 2006.

Smil, Vaclav. *Feeding the World: A Challenge for the Twenty-First Century*. Cambridge, MA: MIT Press, 2000.

Snow, Charles Percy, Sir. *The Two Cultures and the Scientific Revolution*. New York: Cambridge University Press, 1959.

Stent, Gunther. *Paradoxes of Progress*. San Francisco: W.H. Freeman, 1978.

Sulston, John and Georgina Ferry. *The Common Thread: A Story of Science, Politics, Ethics and the Human Genome*. Washington, D.C.: Joseph Henry Press, 2002.

Thomas, Lewis. *The Lives of a Cell: Notes of a Biology Watcher*. New York: Penguin Books, 1978.

van Düring, Monika, Georges Didi-Huberman, and Marta Poggesi. *Encyclopaedia Anatomica: A Complete Collection of Anatomical Waxes*. New York: Taschen, 1999.

BIOGRAPHIES

Editors' Biographies

Sean Caulfield, *University of Alberta*

Sean Caulfield, Canada Research Chair in Printmaking and professor in the Department of Art and Design at the University of Alberta, has exhibited his prints, drawings and book works extensively throughout Canada, the United States, Europe and Japan. Recent exhibitions include: *Inferno* (Open Studio, Toronto, Ontario); *Recent Prints* (Yanagisawa Gallery, Saitama, Japan); and *The 14th International Print Biennial* (Seoul Museum of Art, Korea).

Caulfield has received numerous grants and awards for his work including: SSHRC Fine Arts Creation Grant, Canada Council Travel Grant, and Visual Arts Fellowship (Illinois Arts Council, Illinois, USA). Caulfield's work is in various public and private collections including: Wright State University, Dayton, Ohio; LiuHaisu Art Museum of Shanghai, Shanghai, China; and Purdue University Galleries.

Timothy Caulfield, *University of Alberta*

Timothy Caulfield is research director of the Health Law Institute and a professor in the Faculty of Law and the School of Public Health, University of Alberta. In 2001 he received a Canada Research Chair in Health Law and Policy. He has been involved in a variety of interdisciplinary research endeavours that have allowed him to publish over 150 articles and book chapters. He is a Fellow of the Royal Society of Canada, a senior health scholar with the Alberta Heritage Foundation for Medical Research and the principal investigator on projects funded by Genome Canada, the Stem Cell Network, the Canadian Institutes of Health Research and the Advanced Foods and Materials Network.

Essayists' Biographies

Joan Abrahamson, *Jefferson Institute*

Joan Abrahamson a lawyer, artist and a catalyst for community action, is president of the Jefferson Institute in Los Angeles, a public policy institute that brings creative thinking to practical problems. A major emphasis of the Jefferson Institute is the future of cities. Other areas of activity focus on international security and economics, and health and the study of creativity. Since 1995, she has also been president of the Jonas Salk Foundation. Prior to her work with the Jefferson Institute, Dr. Abrahamson was Assistant Chief of Staff to Vice President George Bush from 1981-1985. She advised him on legal and foreign policy matters. From 1980-1981, she was a White House Fellow, serving as Special Assistant and Associate Counsel to Vice Presidents Walter Mondale and George Bush.

From 1977-1979, Ms. Abrahamson worked with the United Nations Human Rights Commission in Geneva and with UNESCO's Division of Human Rights and Peace, helping to design and implement new procedures for the treatment of alleged violations of human rights.

Lori Andrews, *Chicago-Kent College of Law*

Lori Andrews is a professor at Chicago-Kent College of Law. She graduated *summa cum laude* from Yale College and Yale Law School. She is the author of a ten nonfiction books and three mysteries involving a fictional geneticist, *Sequence*, *The Silent Assassin*, and *Immunity*. She chaired the U.S. Working Group on the Ethical, Legal, and Social Implications of the Human Genome Project and was a consultant to the science ministers of twelve countries. She has advised artists who want to use genetic engineering to become creators with a capital "C" and invent new living species. The *American Bar Association Journal* described Lori as, "a lawyer with a literary bent who has the scientific chops to rival any CSI investigator."

Conrad Brunk, *University of Victoria*

Conrad Brunk is professor of philosophy at the University of Victoria. His areas of research and teaching include ethical aspects of environmental and health risk management, risk perception and communication, and value aspects of science in public policy. Dr. Brunk is a regular consultant to the Canadian government and international organizations on environmental and health risk management and biotechnology. He served as co-chair of the Royal Society of Canada Expert Panel on the Future of Food Biotechnology and, from 2002-2004, as a member of the Canadian Biotechnology Advisory Committee. He is a member of the International Forum for TSE and Food Safety and the Council of Canadian Academies Expert Panel on Nanotechnology. He is co-author of *Value Assumptions in Risk Assessment*, a book exploring how moral and political values influence scientific judgements about technological risks, and author of numerous articles in journals and books on ethical issues in technology, the environment, law and professional practice. Professor Brunk holds a PH.D. in philosophy from Northwestern University.

Cynthia B. Cohen, *Georgetown University*

Cynthia B. Cohen, PH.D., J.D., is Senior Research Fellow at the Kennedy Institute of Ethics at Georgetown University in Washington, D.C. Her publications include eight books, one of which is *Renewing the Stuff of Life: Stem Cells, Ethics and Public Policy* (Oxford, 2007). Dr. Cohen has written and spoken on ethical and policy issues raised by stem cell research, human cloning, the creation of human-non-human chimeras, the new reproductive technologies, end-of-life issues, and religion and bioethics. She has served as a member of the Canadian Stem Cell Oversight Committee, executive director of the National Advisory Board on Ethics in Reproduction, associate legal counsel at the University of Michigan Hospitals, and as consultant to the National Institutes of Health, the National Academies

of Science, the American Association for the Advancement of Science, and the Canadian Stem Cell Network.

Francis S. Collins, *National Institutes of Health*

Francis S. Collins, M.D., PH.D., a physician-geneticist noted for his landmark discoveries of disease genes and his leadership of the Human Genome Project, is director of the National Human Genome Research Institute (NHGRI) at the National Institutes of Health. With Dr. Collins at the helm, the Human Genome Project consistently met projected milestones ahead of schedule and under budget. This remarkable international project culminated in April 2003 with the completion of a finished sequence of the human DNA instruction book. Building on the foundation laid by the Human Genome Project, Dr. Collins is now leading NHGRI's effort to ensure that this new trove of sequence data is translated into powerful tools and thoughtful strategies to advance biological knowledge and improve human health. Dr. Collins is also known for his consistent emphasis on the importance of ethical and legal issues in genetics. In addition to his achievements as the NHGRI director, Dr. Collins's laboratory has discovered a number of important genes, including those responsible for cystic fibrosis, neurofibromatosis, Huntington's disease and, most recently, genes for adult onset diabetes and the gene that causes Hutchinson-Gilford progeria syndrome, a dramatic form of premature aging.

Dr. Collins received a B.SC. from the University of Virginia, a PH.D. in physical chemistry from Yale University, and an M.D. from the University of North Carolina. He has been elected to the Institute of Medicine and the National Academy of Sciences and was awarded the Presidential Medal of Freedom in November 2007.

Edna Einsiedel, *University of Calgary*

Edna Einsiedel is professor of communication studies at the University of Calgary. She is a Genome Canada-funded principal investigator on the project Genomics and Knowledge Translation in Health Systems. Her research interests include social representations of science (focusing on genomics and biotechnology) among publics and stakeholders. She has also examined different forms of public engagement on a variety of genomic applications. She currently serves as editor of the journal *Public Understanding of Science* and is a member of the board of governors for the Council of Canadian Academies of Science.

Jim Evans, *University of North Carolina*
Dr. Evans is professor of genetics and medicine at the University of North Carolina (UNC) and editor-in-chief of *Genetics in Medicine*, the journal of the American College of Medical Genetics. After obtaining his M.D. and PH.D. from the University of Kansas, he served as resident and chief resident of internal medicine at UNC. He trained in medical genetics at the University of Washington.

Dr. Evans is interested in cancer genetics, pharmacogenomics and the broad issue of the use of genetic information. He has served as an advisor and educator to the executive and judicial branches of the U.S. government regarding genetic matters. He lives in Chapel Hill, North Carolina, with his wife, two children, their main dog (Sparky) and auxiliary back-up dog (Lily).

David Garneau, *University of Regina*
David Garneau is Associate Professor of Visual Arts at the University of Regina. He was born in Edmonton and has lived in Regina since 1999. Garneau's practice includes painting, drawing, curation and critical writing. His solo exhibition, *Cowboys and Indians (and Métis?)*, toured Canada (2003–2007). His work often engages issues of nature, history, masculinity and Métis identity. His paintings are in the collections of the Canadian Museum of Civilization, the Canadian Parliament, the Indian and Inuit Art Centre, Glenbow Museum, Mackenzie Art Gallery and many other public and private collections. Garneau has written numerous catalogue essays and reviews and was a co-founder/editor of *Artichoke* and *Cameo* magazines. He is currently exploring the Carlton Trail and roadkill as landscape subjects.

Gail Geller, *Johns Hopkins University*
Gail Geller, SCD, MHS, is a professor at Johns Hopkins University in the Schools of Medicine and Public Health and the Berman Institute of Bioethics. Dr. Geller is an "empirical ethicist" who has earned widespread recognition for her application of the social and behavioural sciences to moral questions in the diffusion of new scientific discoveries in medical care. She is an original member of NHGRI's ethical, legal and social issues (ELSI) research community and has played a leading role in several national grants, training programs, task forces and advisory committees. In particular, Dr. Geller's work has advanced our understanding of patient, provider and public interpretation, communication and decision-making regarding new genetic technologies. Most relevant to this project is her passion for the uncertainties in life and their impact on how we view the world. In her spare time, Dr. Geller exercises her "right brain" through interior design, cooking, writing limericks and travelling.

Hank Greely, *Stanford University*
Hank Greely is the Deane F. and Kate Edelman Johnson Professor of Law and Professor (by courtesy) of Genetics at Stanford University. He specializes in legal and social issues arising from advances in the biosciences. He chairs the California Advisory Committee on Human Stem Cell Research and the steering committee of the Stanford University Center for Biomedical Ethics. He also directs the Stanford Center for Law and the Biosciences and the Stanford Program on Neuroethics. He is one of the founders, and executive committee members, of the Neuroethics Society and is a co-director of the Law and Neuroscience Project, sponsored by the MacArthur Foundation. Professor Greely graduated from Stanford in 1974 and from Yale Law School in 1977.

Jonathan Locke Hart, *University of Alberta*
Jonathan Hart, director of comparative literature, professor of English and associate director of the Centre of Culture & Health, Family Medicine at University of Alberta, is this year a Visiting Fellow, Churchill College,

Cambridge, Northrop Frye Professor at University of Toronto and Invited Professor Universidad de Zaragoza. He has published poetry in literary journals, such as *Harvard Review*, *Grain* and *Mattoid*, for over twenty-five years. His recent collections of poetry include *Breath and Dust* (2000), *Dream China* (2002) and *Dream Salvage* (2003). He has collaborated with Sean Caulfield and Susan Colberg on *Darkfire*, an artist's book, completed in 2007. His recent scholarly books include *Representing the New World* (2001), *Comparing Empires* (2003), *Columbus, Shakespeare and the Interpretation of the New World* (2003), *Contesting Empires* (2005), *Interpreting Cultures* (2006) and *Empires and Colonies* (2008).

Anna R. Hayden, *McGill University*

Anna Hayden has just completed her combined arts and science degree at McGill University, where she graduated with great distinction. Anna has spent a large amount of time pursuing the interface between arts and science in her role as reporter on science for the McGill student newspaper, as well as in her volunteer activities for different organizations. She has worked in research laboratories in Vancouver, Montreal and Amsterdam. Anna is currently volunteering in Argentina with an AIDS organization. She now wishes to continue the integration of humanities and science with her pursuit of the study of medicine.

Michael R. Hayden, *University of British Columbia*

Dr. Michael Hayden is a University Killam Professor at the University of British Columbia, as well as director of the Centre for Molecular Medicine and Therapeutics (CMMT), a gene research centre under UBC's Faculty of Medicine. Author of over six hundred peer-reviewed publications and invited submissions, Dr. Hayden focusses his research primarily on genetic diseases, including genetics of lipoprotein disorders, Huntington's disease and predictive medicine.

Dr. Hayden is the recipient of numerous prestigious honours and awards, which include the Lifetime Achievement award by the Huntington Society of Canada in 2001, the Leadership and Research Excellence award by the National Centres of Excellence in 2004 and the BC Biotech Life Sciences Company of the Year in 2005. His latest award includes the Prix Galien 2007.

Jay Ingram, *Discovery Channel*

Jay Ingram has been the host of Discovery Channel Canada's *Daily Planet* since it began in 1995. At the time it was the only hour-long, prime-time daily science show. Prior to joining Discovery, Jay hosted CBC Radio's national science show, *Quirks and Quarks*, from 1979 to 1992. During that time he won two ACTRA awards, one for best host. He was a contributing editor to *Owl* magazine for ten years and wrote a weekly science column in the *Toronto Star* for twelve. Jay has written ten books and has another on the way.

He has received the Sandford Fleming medal from the Royal Canadian Institute for his efforts to popularize science, the Royal Society's McNeil medal for the Public Awareness of Science, as well as the Michael Smith award from the Natural Sciences and Engineering Research Council. He is a Distinguished Alumnus of the University of Alberta and has received four honorary doctorates.

Bartha Maria Knoppers, *Université de Montréal*

Bartha Maria Knoppers, PH.D., holds the Canada Research Chair in Law and Medicine and the Chaire d'excellence Pierre Fermat (France). She is Professor at the Faculté de droit, Université de Montréal and senior researcher at the Centre for Public Law. Dr. Knoppers is former chair of the International Ethics Committee of the Human Genome Organization (1996-2004) and member of UNESCO's International Bioethics Committee, which drafted the Universal Declaration on the Human Genome and Human Rights (1993-1997). She is co-founder of the International Institute of Research in Ethics and Biomedicine and chair of the Ethics Working Party of the International Stem Cell Forum. In 2003, she founded the international Public Population Project in Genomics (P3G).

Trudo Lemmens, *University of Toronto*

Trudo Lemmens is associate professor in the Faculties of Law and Medicine of the University of Toronto. He was a visiting professor at Otago University (New Zealand) the K. U. Leuven (Belgium), a member of the Institute for Advanced Study in Princeton, and a Fellow of the Royal Academy of Belgium for Science and the Arts. He is particularly interested in how law can contribute to the promotion of ethics in medical research and in the development of new healthcare technologies. His publications include the co-authored book *Reading the Future? Legal and Ethical Challenges of Predictive Genetic Testing* (Thémis, 2007) and the co-edited book *Law and Ethics of Medical Research: Regulation, Conflict of Interest and Liability* (University of Toronto Press, 2006).

Lianne McTavish, *University of Alberta*

Lianne McTavish, PH.D., is professor of the history of art, design, and visual culture at the University of Alberta. Her SSHRCC-funded research on early modern medical imagery has produced articles for the *Social History of Medicine*, *Medical History* and a monograph, *Childbirth and the Display of Authority in Early Modern France*. Her recent work in this area analyzes representations of cure and convalescence in France, 1600–1800. Lianne has also published on the history and theory of museums in *Cultural Studies*, *Acadiensis*, *New Museum Theory and Practice*, *Canadian Historical Review* and the *Journal of Canadian Studies*. She is currently completing a book manuscript called *Between Museums: Exchanging Objects, Values and Identities, 1842–1950*.

Eric M. Meslin, *Indiana University*

Eric M. Meslin, PH.D., is director of the Indiana University Center for Bioethics, associate dean and professor of medicine, medical and molecular genetics and philosophy. Eric has a B.A. in philosophy from York University in Toronto and both an M.A. and PH.D. from the Kennedy Institute of Ethics at Georgetown University. He has held academic positions at the University of Toronto and at the University of Oxford. His scientific imagination was stimulated by working on the Human Genome Project at the National Human Genome Research Institute and then at the White House for President Clinton's National Bioethics Advisory Commission. While he dabbled in art as a youngster, his artistic imagination now comes solely from regular visits to museums and galleries.

Peter W.B. Phillips, *University of Saskatchewan*

Dr. Peter W.B. Phillips, an international political economist, is professor and head of political studies, acting director of the new School of Public Policy and an associate member of the Departments of Economics, Bioresource Policy, Business and Economics and Management at the University of Saskatchewan. He holds a concurrent appointment as Professor at Large at the Institute for Advanced Studies, University of Western Australia, Perth. His research concentrates on issues related to governing transformative innovation. He is the CO-PI of the Genome Alberta project on Translating Knowledge in Health Systems (2006–2010) and a collaborator on six other internationally peer-reviewed research programs. His latest book, *Governing Transformative Technological Innovation: Who's in charge?*, was published by Edward Elgar in June 2007.

Jai Shah, *Harvard University*

Jai Shah's research has focused on the developing interface between pharmacogenomics and economic, regulatory and healthcare environments and forces. He remains interested in this area as part of a broader curiosity regarding advances in genomics and molecular medicine, their relevance to social, policy and public health issues, and their uptake within and across health systems. Jai holds degrees in biological sciences, health policy, and clinical medicine, has spent time at the Nuffield Council on Bioethics in London, UK, and contributed to research policy at the Canadian Institute of Health Research in Ottawa. He is currently a resident in psychiatry at Harvard University's Longwood Program in Boston, Massachusetts.

Stephen Strauss, *CBC*

Stephen Strauss was a science writer for over twenty years with the *Globe and Mail* and more recently has written a column for the CBC's website. He has won six Canadian Science Writer's Science In Society awards, was the first winner of the Connaught Medal for Medical Reporting and, in 2007, won the UBC Graduate School of Journalism's Prize for Internet Science Journalism and then the Barry Lando Prize for Best Science Journalism Overall in print, broadcasting and on the Internet. He has authored several book chapters and three books—one of which was a children's book on measurement written all in rhyme. In 2007-2008 he was writer-in-residence at McMaster University's Arts and Science Program.

Michele Veeman, *University of Alberta*

Michele Veeman is professor emerita of agricultural and resource economics, Department of Rural Economy, University of Alberta, Canada. She holds a PH.D. in agricultural economics (University of California, Berkeley), a master's degree in economics (University of Adelaide, South Australia) and a bachelor's degree in agricultural science (Massey University, New Zealand). Her research and teaching have focused on the economics of food, agriculture and rural resources; she has published widely in these areas. Continuing research includes studies of individuals' risk perceptions, decisions and trade-offs relative to food biotechnology. She is a Distinguished Scholar of the Western Association of Agricultural Economics, a Fellow of the Canadian Agricultural Economics Society and an Honorary Life Member of the International Association of Agricultural Economists.

Gregor Wolbring, *University of Calgary*

Gregor Wolbring, PH.D., is a biochemist, a bioethicist, as well as an ability studies, health policy and sociology of nano-, bio-, info-, cogno- (Neuro-engineering) synthetic biology researcher. He is an assistant professor in the Faculty of Medicine, University of Calgary, Canada. He is a founding member and affiliated scholar at the Center for Nanotechnology and Society at Arizona State University,

USA, a member of the International Nanotechnology and Society Network, a member of Canadian Advisory Committees for the International Organization for Standardization section TC229 Nanotechnologies (CAC/ISO), and part-time professor, Faculty of Law, University of Ottawa, Canada, and adjunct professor at the Faculty of Critical Disability Studies, York University, Canada. His webpage is www.bioethicsanddisability.org, and his biweekly column, *The Choice is Yours,* can be found here at: http://www.innovationwatch.com/commentary_choiceisyours.htm. His blog is at www.wolbring.wordpress.com.

Artists' Biographies

Fiona Annis, *Glasgow School of Art*

Fiona Annis is a practicing visual artist and researcher at the Glasgow School of Art. Based between Glasgow and Montréal, her practice includes documentation, installation, print media and curation, with a focus on the notions of co-culture, the control inherent in care, and an examination of the body(ies) in the contemporary context. In 2007, Fiona co-curated *Post-Amen* in collaboration with Jamie Ferguson, as well as *Assemble: a gathering of performative gestures,* at the Assembly Gallery of Glasgow School of Art.

Fiona is currently exploring a new cycle of work, *Corpscellule | 6616,* in collaboration with JPC/K, also known as Shawn Bailey. *Corpscellule* has recently exhibited at *Intimacy: Across Digital & Visceral Performance Platform,* Goldsmith's University of London, as well as Lowsalt Gallery, in the context of the Gi, the Glasgow International Festival of Contemporary Visual Art. *Corpscellule* is an ongoing cycle of work that explores the intersections of the sublime and the grotesque, unravelling the peripheries of revelation and concealment, employing the body variously as material, site and agent.

Shawn Bailey, *Concordia University*

Shawn Bailey is a practicing artist working with digital print media, video and installation. His current research explores notions of authority, control structures, media and international biotech and pharmaceutical policies. Since 2002, Bailey and Jennifer Willet have collaborated on an innovative computational, biological, artistic project called *BIOTEKNICA*. *BIOTEKNICA* has been exhibited at ISEA San Jose, USA, in collaboration with Oron Catts and Ionat Zurr's *Tissue Culture and Art Project* (2006), *Biennial Electronic Arts Perth* (Australia, 2004), *The European Media Arts Festival* (Germany, 2003), *La Société des arts et technologiques* (SAT) (Canada, 2005) and *The Forest City Gallery* (Canada, 2004).

Christine Borland, *Glasgow School of Art*
Christine Borland is based in Argyll, Scotland; she studied at Glasgow School of Art and the University of Ulster, Belfast. Her work is associated with the systems and processes that underpin society—both current and archaic—including forensic science, medicine and biotechnology. These intersections are revealed in a spectrum of projects ranging from gallery installations to book works and public sculpture projects. She is in the middle of a three-year NESTA Fellowship exploring the incorporation of humanities into medical education; she is also an academic researcher at Glasgow School of Art. Her work has been exhibited extensively internationally—her recent solo show, *Preserves*, at the Fruitmarket Gallery, Edinburgh, is accompanied by a publication documenting works of the last fifteen years. In 1997, she was nominated for the Turner Prize. Christine Borland is represented by Lisson Gallery, London, Sean Kelly Gallery, New York, Galeria Toni Tapies, Barcelona, and Cent 8, Paris.

Sean Caulfield, *University of Alberta*
Sean Caulfield, Canada Research Chair in Printmaking and professor in the Department of Art and Design at the University of Alberta, has exhibited his prints, drawings and book works extensively throughout Canada, the United States, Europe and Japan. Recent exhibitions include: *Inferno* (Open Studio, Toronto, Ontario); *Recent Prints* (Yanagisawa Gallery, Saitama, Japan); and *The 14th International Print Biennial* (Seoul Museum of Art, Korea).

Caulfield has received numerous grants and awards for his work including: SSHRC Fine Arts Creation Grant, Canada Council Travel Grant, and Visual Arts Fellowship (Illinois Arts Council, Illinois, USA). Caulfield's work is in various public and private collections including: Wright State University, Dayton, Ohio; LiuHaisu Art Museum of Shanghai, Shanghai, China; and Purdue University Galleries.

Christine Davis, *Toronto*
Since 1984 Christine Davis has had an extremely active professional career, participating in solo and group exhibitions across Canada, the United States and Europe. Her creative practice includes a variety of media including photograph, time-based works and installation. Her creative work has also involved utilizing the body as subject matter and in installations Davis has combined artifacts from the body, such as genetic coding sequences, with other images to poetically examine contemporary culture's constructions of self. Her compelling work has been acquired by the permanent collection of numerous important public institutions including: Art Gallery of Ontario, Toronto, Ontario; Bibliothèque nationale de France, Paris; Canada Council Art Bank, Ottawa, Ontario; and Canadian Museum of Contemporary Photography, Ottawa, Ontario.

Bernd Hildebrandt, *University of Alberta*
Communication through images and type are central to Bernd Hildebrandt's work. With an interest in photography and sculpture, his master's degree from the University of Alberta dealt with the structure, material and perception of words and letters in an environment. As an exhibit designer, Bernd has done numerous exhibits on art, as well as mummification, Chinese robes, dentistry, diamonds, meteorites and biographical material. His interest in an exhibit's interpretation of the object, and utilization of space, has also influenced his personal work, which includes solo and group installations of natural materials, poetic texts and images. Currently working as a freelance designer, Bernd teaches classes at the University

of Alberta in integrative design, which combine two- and three-dimensional design disciplines.

Liz Ingram, *University of Alberta*

Liz Ingram was born in Argentina and grew up in New Delhi, Mumbai and Toronto. She is professor in print-making at the University of Alberta. She has exhibited in over twenty solo and duo exhibitions, and over two hundred group exhibitions in North and South America, Europe, the Middle East and the Far East and has received awards at exhibitions in Canada, Slovenia, Korea, Brazil, Estonia, India and Finland. In 2008, she received the University of Alberta J. Gordon Kaplan Award for Research Excellence. She has been visiting artist at many universities including the Tokyo National University, Musashino University (Tokyo) and the University of Applied Sciences (Münster, Germany). The human body and the elemental in nature are recurring subjects in her art practice.

Eduardo Kac, *The School of the Art Institute of Chicago*

Eduardo Kac is internationally recognized for his inter-active telepresence works and his bio art. A pioneer of telecommunications art in the pre-Web 1980s, Eduardo Kac (pronounced "Katz") emerged in the early 1990s with his radical telepresence and biotelematic works. His visionary combination of robotics and networking explores the fluidity of subject positions in the post-digital world. At the dawn of the twenty-first century, Kac opened a new direction for contemporary art with his "transgenic art"—first with a groundbreaking piece entitled *Genesis* (1999), which included an "artist's gene" he invented, and then with "GFP Bunny," his fluorescent rabbit called Alba (2000). Kac's work has been exhibited internationally at venues such as Exit Art and Ronald Feldman Fine Arts, New York; Maison Européenne de la Photographie, Paris, and Lieu Unique, Nantes, France; OK Contemporary Art Center, Linz, Austria; InterCommunication Center (ICC), Tokyo; Julia Friedman Gallery, Chicago; Seoul Museum of Art, Korea; and Museum of Modern Art, Rio de Janeiro. Kac's work has been showcased in biennials such as Yokohama Triennial, Japan, Gwangju Biennale, Korea, and

Bienal de Sao Paulo, Brazil. His work is part of the perma-nent collection of the Museum of Modern Art in New York, the Museum of Modern Art of Valencia, Spain, the ZKM Museum at Karlsruhe, Germany, and the Museum of Modern Art in Rio de Janeiro, among others. His work is documented on the Web at http://www.ekac.org.

Royden Mills, *University of Alberta*

Royden Mills is a grandson of an original Canadian prairie homesteader. He received a diploma in architecture from the Northern Alberta Institute of Technology. Art and Design studio studies began at Red Deer College and culminated with a master's degree from the University of Alberta. Mills worked closely with Anthony Caro as assistant in 1989. Mills subsequently travelled to international workshops in Poland and the USA and maintained a studio in Hokkaido, Japan for two years. He has been coordinator of thirty-two sections of art and design fundamentals at the University of Alberta. Mills's sculptures have been exhibited and included in collections internationally, including at Arlington Heights Sculpture Park (Chicago), Convergence Festival (Providence, Rhode Island), Centennial Plaza (Red Deer, Canada), Franconia Sculpture Park (USA), Windsor Sculpture Park (State University of New York), Chomin Hall (Hokkaido, Japan), Collection of Alberta Foundation for the Arts (MacEwan College) and the University of Lethbridge.

Lyndal Osborne, *University of Alberta*

Lyndal Osborne has worked in a wide range of media including print, drawing, sculpture and installation. She was born and completed undergraduate education in Australia. She completed an MFA from the University of Wisconsin, Madison, USA. Currently she is professor emer-itus at the University of Alberta, Edmonton. Her work has been exhibited in Canada, Europe and Asia. She has received numerous awards and grants. Lyndal is repre-sented in private and public collections including the National Gallery of Canada. The focus of her work in both printmaking and installation has been informed by a long-term fascination with the natural world. Collecting and categorizing, vulnerability and power, preservation and endangerment are all areas examined in her work.

Catherine Richards, *University of Ottawa*

Catherine Richards is a visual artist working in old and new media art. Her work explores the volatile sense of ourselves as we are shifting our boundaries—a process in which new information technologies play a staring role.

The Canada Council for the Arts awarded the Media Arts prize (Petro-Canada, 1993) to her work *Spectral Bodies*, whose use of virtual reality technology was an "outstanding and innovative use of new technologies in media arts." She received a Canadian Centre for the Visual Arts Fellowship (1993-94) at the National Gallery of Canada, which subsequently commissioned her piece *Charged Hearts*. She co-directed BIOAPPARATUS, an interdisciplinary residency on art, intimacy and new technologies at the Banff Centre .

She considers new technologies as art material. Her work explores the spectator's role in these technologies like "jam in the electro-magnetic sandwich."

Her works have been supported by the Canada Council for the Arts; AT&T Canada; The Claudia De Hueck Fellowship in Art and Science at the National Gallery of Canada; the Daniel Langlois Foundation for Art, Science and Technology, the University of Ottawa; as well as the contributions of individual scientists. Recently, Catherine Richards received the AIRes fellowship—an artist-in-residence position at the National Research Council of Canada. Currently she is an associate professor in the Department of Visual Arts at the University of Ottawa.

Jennifer Willet, *Concordia University*

Jennifer Willet is a PH.D. candidate in the Interdisciplinary Humanities program at Concordia University. Her work explores notions of self and subjectivity in relation to biomedical, bioinformatics and digital technologies. Since 2002, Willet and Shawn Bailey have collaborated on an innovative computational, biological, artistic, project called *BIOTEKNICA*. *BIOTEKNICA* has been exhibited in various forms including ISEA San Jose, USA,

in collaboration with Oron Catts and Ionat Zurr's *Tissue Culture and Art Project* (2006), *Biennial Electronic Arts Perth* (Australia, 2004), *The European Media Arts Festival* (Germany, 2003), *La Société des arts et technologiques* (SAT) (Canada, 2005), and *The Forest City Gallery* (Canada, 2004).

Adam Zaretsky, *Rensselar Polytechnic Institute*

Adam Zaretsky is a Vivoartist working in biology and Art Wet Lab Practice. This involves biological lab immersion as a process toward inspired artistic projects. His personal research interests revolve around life, living systems, exploration into the mysteries of life and interrogating varied cultural definitions that stratify life's popular categorizations. He also focusses on legal, ethical and social implications of some of the newer biotechnological materials and methods: molecular biology, assisted reproductive technology (ART) and transgenic protocols. Zaretsky also teaches Vivoarts: Ecology, Biotechnology, Non-human Relations, Live Art and Gastronomy. Focus is on artistic uses and the social implications of molecular biology, tissue culture, genomics and developmental biology.